Logistics Transportation Systems

Quick Guide to the Text

This text is organized by lecture modules of IET 470/570 course at the University of Southern Mississippi. It has a total of 12 lecture modules. Each module consists of Power Point handouts and supporting reading materials. Power Point handouts provide a guideline of discussion topics covered in the course and supporting reading materials elaborate the concepts related to the lecture modules. The text covered following 12 modules:

1) Logistics transportation fundamentals
2) Transportation costs analysis
3) Transportation networks analysis
4) Transportation infrastructure & equipment
5) Transportation rules & regulations
6) Intermodal transportation system
7) Logistics transportation problems with linear programming
8) Assignment and transportation problems with linear programming
9) Logistics customer services
10) Transportation rates analysis
11) Transportation decisions analysis
12) Transportation security & military logistics

Logistics Transportation Systems

First Edition

MD Sarder, PhD.

University of Southern Mississippi

Lulu Press Inc.

Raleigh, NC, USA

Table of Contents

Preface

I have developed the IET 470/570: Logistics Transportation Systems course and I have been teaching this course since 2009. In spite of my best efforts to make it enjoyable and interesting, I found many of my students struggle in this course. Some of them might have blamed me for my shortcomings as an instructor. But, all along I felt that the real problem has been the textbook. I considered other texts too, and found them equally unsatisfactory, primarily due to the lack of coverage of all topic areas and the overemphasis on engineering science. Logistics transportation is an evolving area and no single text book covers all the areas that are needed for today's graduates.

This text is basically a course guideline for students in Logistics Transportation Systems course where students can get the lecture notes and elaborate discussions on topics related to lecture modules. This text is an intermediate product of a formal text book on Logistics Transportation Systems, which is in the process of publishing and will be available soon.

LECTURE I

POWER POINT HANDOUTS

Logistics Transportation Overview

Dr. MD Sarder

Logistics Transportation

It is the movement of goods efficiently from the origin to the destination.

Efficiently means –
- Right place/person
- Right time
- Right product
- Right condition &
- Right price

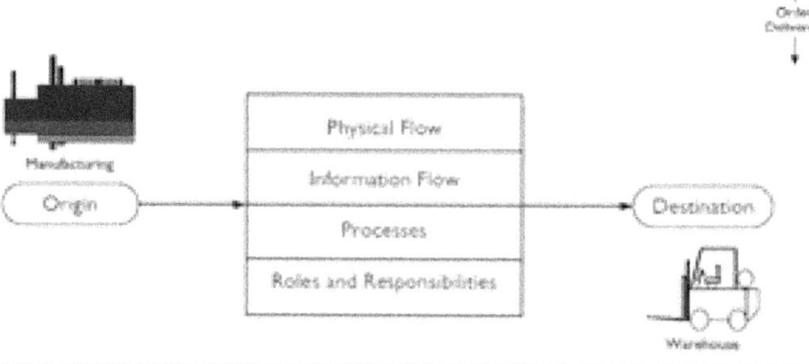

Transportation Management

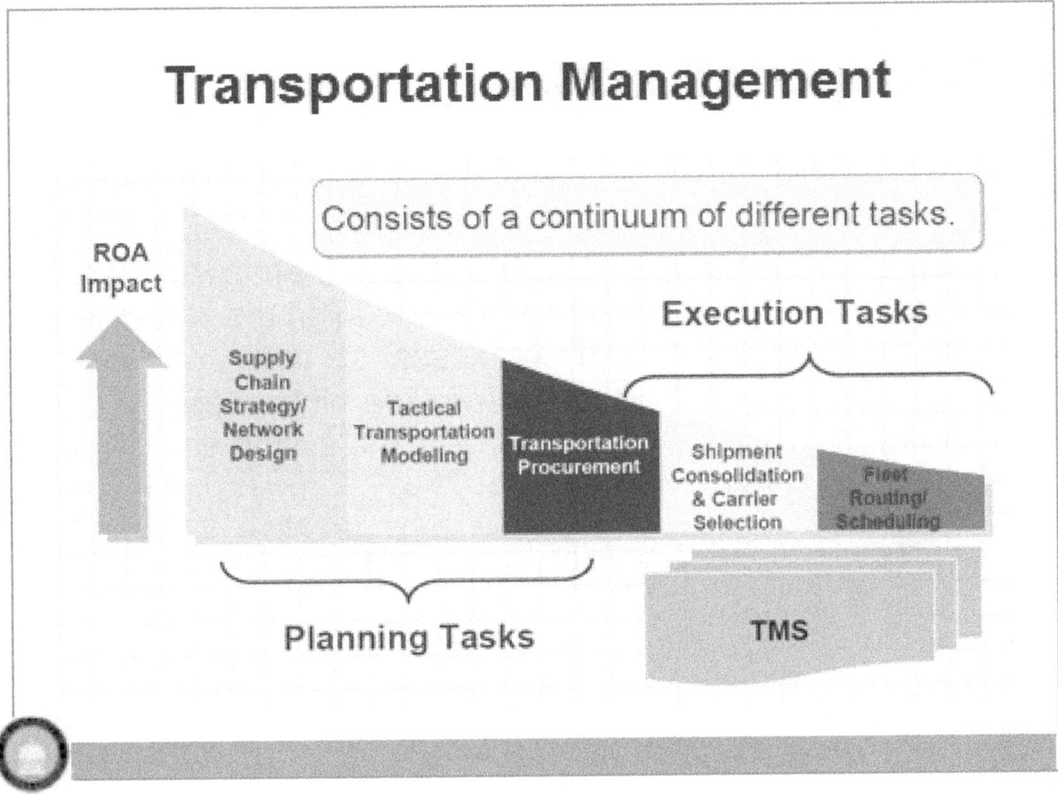

Transportation Management

Strategic	· What carriers should I partner with and how? · How will seasonality affect my carrier assignments? · Should I use dedicated or private fleets? · Which carriers provided quality service in the past? · Should I use pool points, cross-docks, or multi-stop routes?
Tactical	· How can I quickly secure rates for a new DC/plant/lane? · What lanes are having performance problems? · Which carriers are complying to or exceeding their contracts? · Are site managers are complying to the strategic plan? · Where should I establish a seasonal contract?
Operational	· Which carrier should I tender this load to? · How can I collaboratively source this weeks' loads? · How do I prevent Maverick/Rogue behavior? · Should I use a contract carrier or look at the spot market? · How can I best communicate with my carriers?

Transportation Management Layers

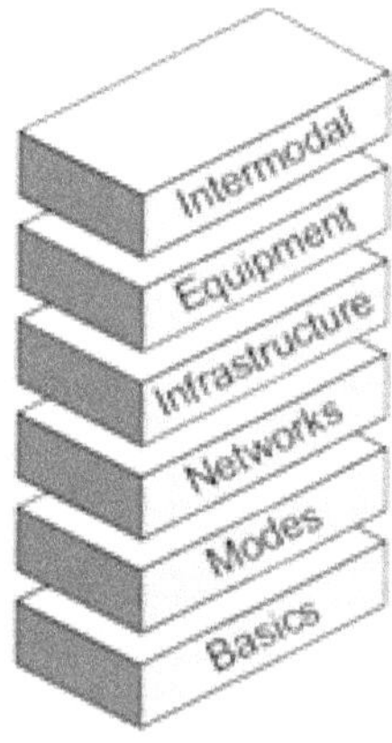

➢ Transportation consists of many layers.

➢ Transportation managers find the best ways to align the strength and weakness of these various layers to provide high quality, low cost service to the customers.

Transportation Processes

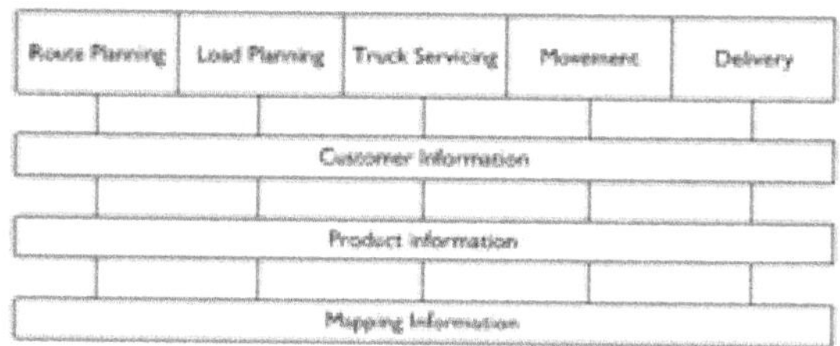

Transportation Processes consists of

1. Route Planning establishes the route the truck will take to make its deliveries. The trade-off is one of time and expense against commitments to the customer.
2. Load Planning is arranging the shipments within the truck to minimize handling. For example, one might organize on a first-off last-loaded model.
3. Truck Servicing assures that the truck is fueled and otherwise properly prepared for the trip.
4. Movement is the physical movement of the truck along the planned route.
5. Delivery puts the product in the buyers hands. This would require, in the case of this LCD television, a signature (delivery receipt) from the buyer.

Transportation Cost

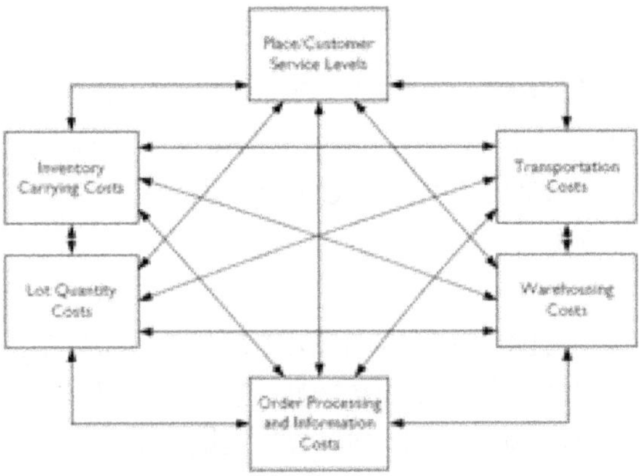

Transportation as a component of cost

➢Transportation is but one of the components of cost that needs to be considered in providing the level of service desired by the customer at a price the customer is willing to pay.

Transportation Cost

Transportation Costs Impact on →	Warehousing Costs	Order Processing and Information Costs	Lot Quantity Costs	Inventory Carrying Costs
Higher	Lower; less product is required in the warehouse to maintain the desired service level.	Higher; these costs are a function of order size. Higher transportation costs implies faster transportation in turn implying smaller order sizes.	Higher; these costs are a function of manufacturing volume. Higher transportation costs implies faster transportation in turn implying smaller manufacturing lots.	Lower; less product is required in the warehouse to maintain the desired service level.
Lower	Higher; more product is required in the warehouse to maintain the desired service level.	Lower; these costs are a function of order size. Lower transportation costs implies slower transportation in turn implying larger order sizes.	Lower; these costs are a function of manufacturing volume. Lower transportation costs implies slower transportation in turn implying larger manufacturing volumes.	Higher; more product is required in the warehouse to maintain the desired service level.

Transportation Cost

➢One should not, therefore, think that the decision as to which transportation mode to use is simply based on transportation costs alone. It is more complex than that.

➢This suggests that transportation management, as meant to be the selection and management of transportation service, is critically dependent upon an understanding of the business of the customer.

Transportation Options
Modes of Transportation

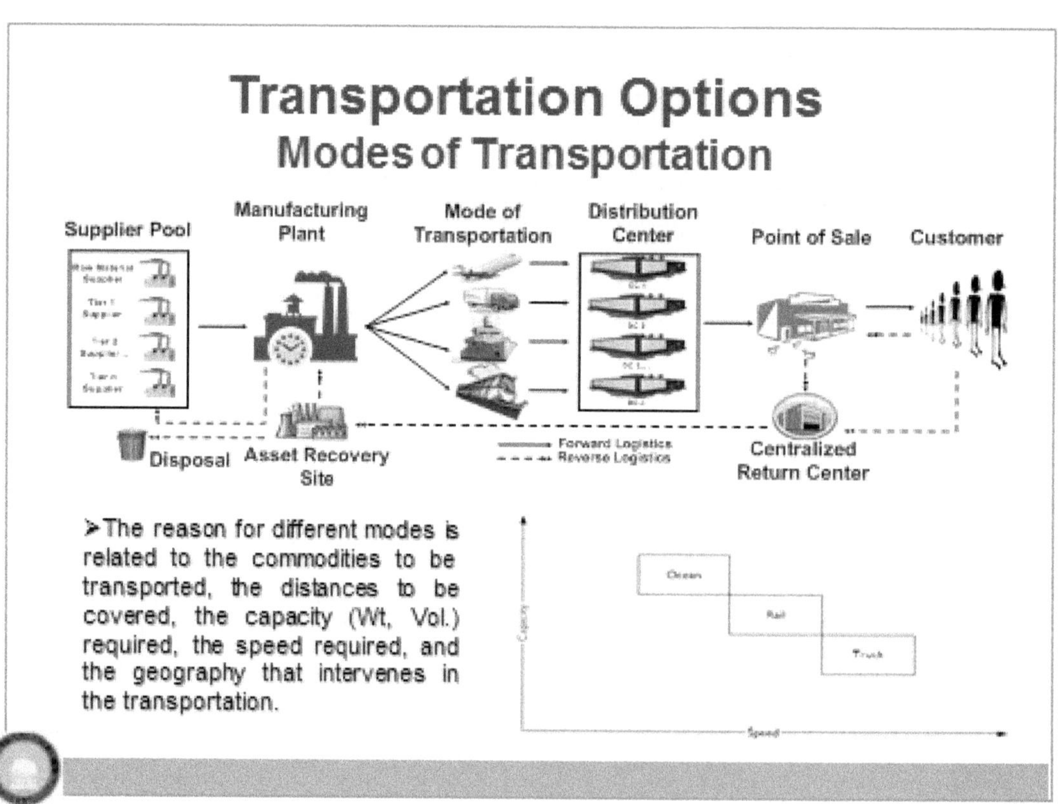

➢The reason for different modes is related to the commodities to be transported, the distances to be covered, the capacity (Wt, Vol.) required, the speed required, and the geography that intervenes in the transportation.

Transportation Options
Modes Selection

◆ When is it better to use a cheaper more variable transport mode?

- Air – higher v, smaller σ
- Rail – lower v, larger σ

$$\Delta TRC = TRC_a - TRC_r$$

where,

$$TRC_a = \sqrt{2A_a Dv_a r} + k_a \sigma_{L,a} v_a r + Dv_a$$

$$TRC_r = \sqrt{2A_r Dv_r r} + k_r \sigma_{L,r} v_r r + Dv_r$$

◆ Pick mode with smaller TRC

Transportation Options
Modes Selection

◆ Multiple Modes

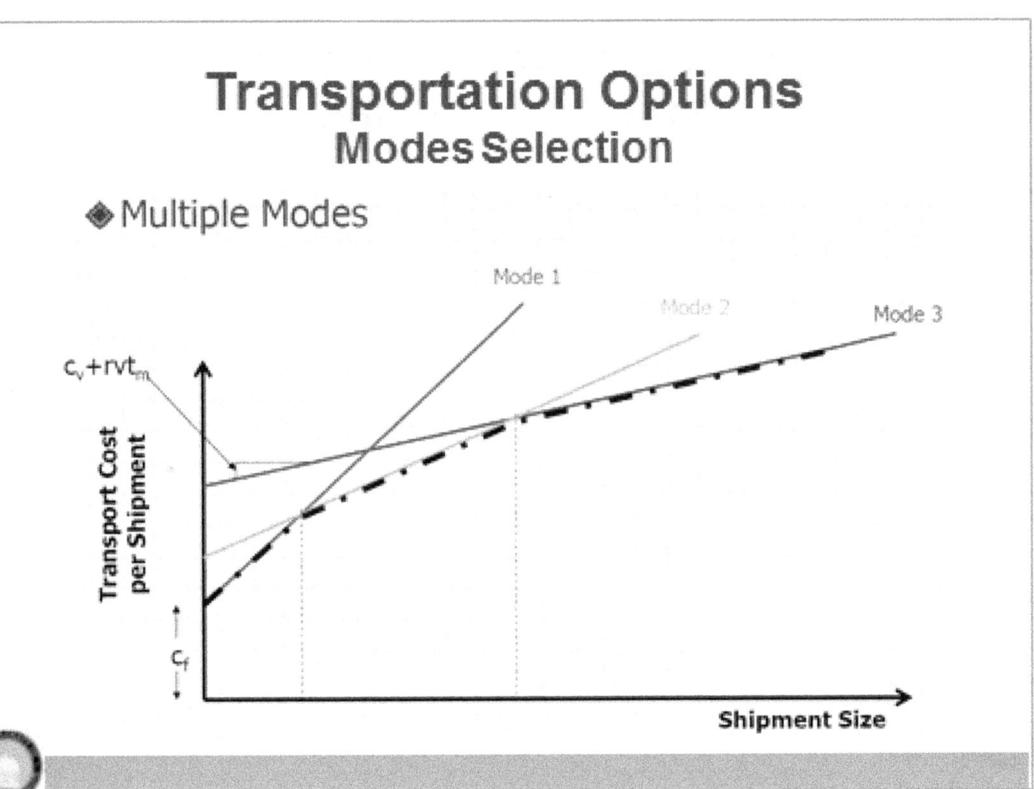

Transportation Options
Modes Selection

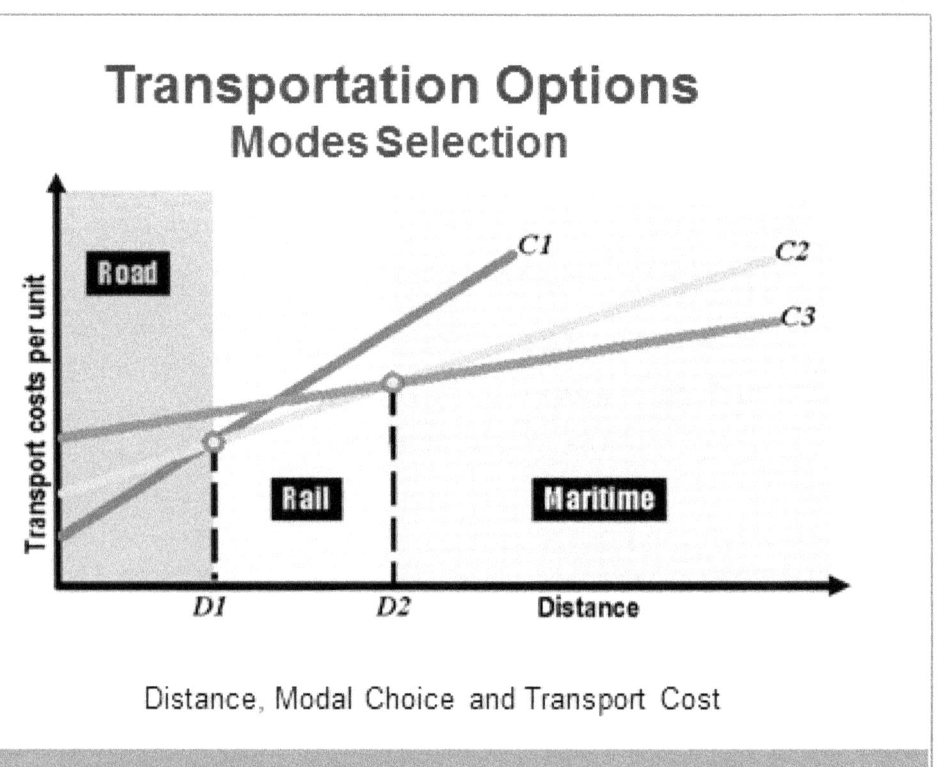

Distance, Modal Choice and Transport Cost

Transportation Options
Moving vs. Storing

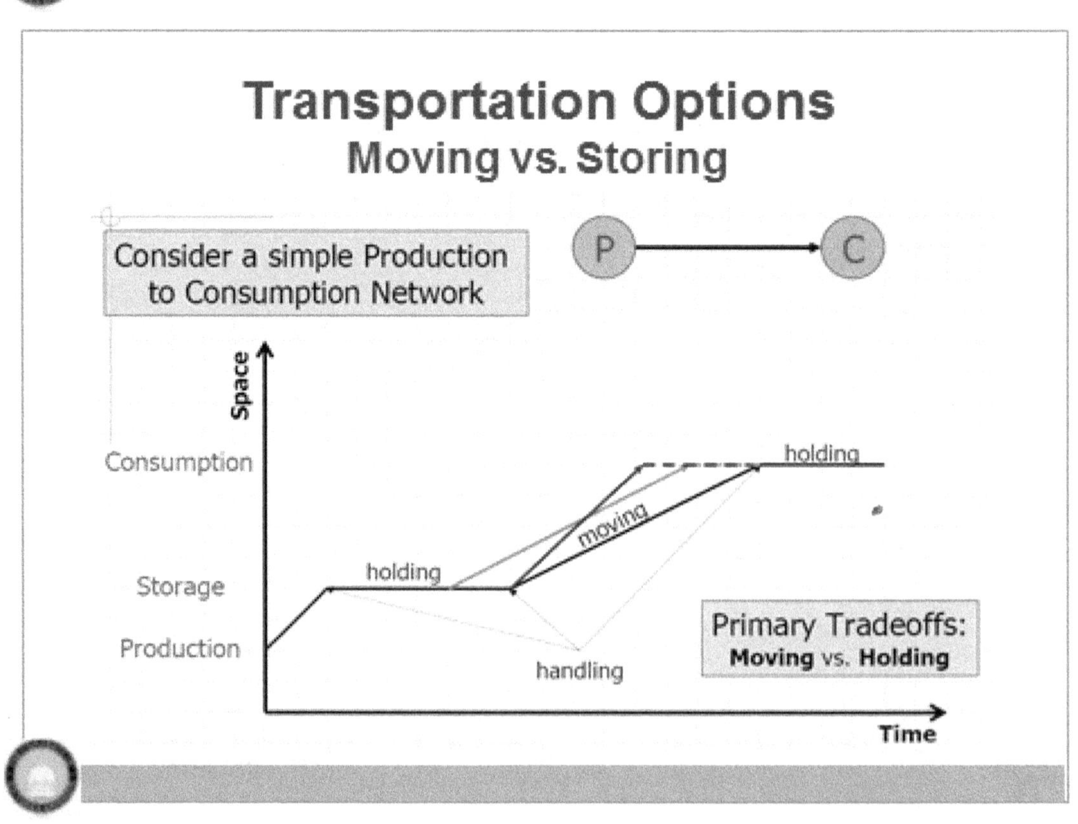

Transportation Options
Modes Comparison

Identify at least 12 characteristics of following modes

- Air--

- Truck--

- Rail--

- Water--

- Pipe--

Transportation Options
Modes Comparison

	Truck	Rail	Air	Water
Operational Cost	Moderate	Low	High	Low
Market Coverage	Pt to Pt	Terminal to Terminal	Terminal to Terminal	Terminal to Terminal
Degree of competition	Many	Few	Moderate	Few
Traffic Type	All Types	Low to Mod Value, Mod to High density	High value, Low density	Low value, High density
Length of haul	Short – Long	Medium – Long	Long	Med - Long
Capacity (tons)	10 – 25	50 – 12,000	5 – 12	1,000 – 6,000

Transportation Options
Modes Comparison

	Truck	Rail	Air	Water
Speed	Moderate	Slow	Fast	Slow
Availability	High	Moderate	Moderate	Low
Consistency (delivery time)	High	Moderate	Moderate	Low
Loss & Damage	Low	High	Low	Moderate
Flexibility	High	Low	Moderate	Low

	Truck	Rail	Air	Water	Pipeline
BTU/ Ton-Mile	2,800	670	42,000	680	490
Cents / Ton-Mile	7.50	1.40	21.90	0.30	0.27
Avg Length of Haul	300	500	1000	1000	300
Avg Speed (MPH)	40	20	400	10	5

LECTURE I
READING MATERIALS

1.0 INTRODUCTION

In the current state of our global market, the logistics manager acts in a pivotal role for companies as they bring their product from source to market. For many companies, the transportation of goods from production to end user constitutes up to two thirds of total costs associated with a particular item. Since the movement of goods through the supply, or logistics, plays a critical role in determining the overall cost of delivering a product to market, it generates significant revenues for companies and firms around the globe. According to the North American Transportation Statistics Database, more than three trillion metric tons of freight was transported domestically throughout the United States in 2010. The massive shipments of freight represent a large portion of the global economy. Freight shipments in how the world moves its money. "On a typical day in 2007, over 35.7 million tons of goods, valued at $32.4 billion, moved nearly 9.6 billion ton-miles on the nation's transportation network (US Dept. of Transportation, 2008)." Of these freight shipments more than 90% where shipped using a single mode of transportation and the remaining percentage were shipped using two of more modes of transportation (US Dept. of Transportation, 2008). In 2010, the United States exported trillions worth of merchandise all over the world (US Dept. of Transportation, 2009). The transportation and transportation related service industries employed 12 million Americans alone in 2009, which constitutes 9.3% of the total labor pool (US Dept. of Transporation, 2010) These numbers are staggering but not surprising when you consider how diversity of the logistics field. There are five major modes of transportation, Air, Rail, Road, maritime, and pipeline, and each of these modes of transportation must have a number of employees directly working on transportation operations and additional employees of service companies that provide other services directly related to these same operations, such as maintenance. These logistics service personnel play a pivotal role in ensuring that that fleet operations are efficient. This can be a bit complicated when you consider that in 2009 there were more than one million freight cars and locomotives as well more than 40,000 barges or ocean going ships utilizing the U.S. transportation system (US Dept. of Transporation, 2010). These numbers only consider the U.S. transportation system. If you expand the scope to the global logistics transportation system you will begin to understand why logistics management is a growing field. The management and improvement of global logistics transportation systems will be a challenge for future logisticians because we must improve the efficiency of our given system while of the same time ensuring that growth is managed properly and effectively. Table 1 provides a sampling of the physical infrastructure found around the world in 15 countries.

With transportation playing such a crucial role in business operations logisticians are asked to serve in two main capacities. First, Logisticians expected to act as subject matter experts on transportation systems design and components within a given supply chain. Second, the logistician must act as a catalyst for improvement in the operating systems and decision making associated with supply chain management. This text will serve to broaden the logistician's knowledge on transportation systems, specifically key terms, components of the supply chain, and modes of transportation. Additionally, the chapter will cover pertinent transportation industry rules and regulations. To lay the groundwork for our study of logistics transportation systems let us begin with a definition.

Table I: Extent of the Physical Transportation Systems in World's Top Economies: 2008

Ranked by total roadways	Roadways		Railways	Waterways	Pipelines	Airports
	Total (km)	Paved roads (km)	(km)	(km)	(km)	(number)
United States	6,465,799	4,209,835	226,427	41,009	793,285	5,146
India	3,316,452	1,517,077	63,327	14,500	22,773	251
China	1,930,544	1,575,571	77,834	110,000	58,082	413
Brazil	1,751,868	96,353	28,857	50,000	19,289	734
Japan	1,196,999	949,101	23,506	1,770	4,082	144
Canada	1,042,300	415,600	46,688	636	98,544	514
France	951,500	951,500	29,213	8,501	22,804	295
Russia	933,000	754,984	87,157	102,000	246,855	596
Australia	812,972	341,448	37,855	2,000	30,604	462
Spain	681,224	681,224	15,288	1,000	11,743	154
Germany	644,480	644,480	41,896	7,467	31,586	331
Italy	487,700	487,700	19,729	2,400	18,785	101
Turkey	426,951	177,500	8,697	1,200	11,191	103
Sweden	425,300	139,300	11,633	2,052	786	249
Poland	423,997	295,356	22,314	3,997	15,792	126
United Kingdom	398,366	398,366	16,454	3,200	12,759	312
Indonesia	391,009	216,714	8,529	21,579	13,752	669
Mexico	356,945	178,473	17,516	2,900	40,016	243

(US Dept. of Transporation , 2009)

There are many accepted definitions of the word logistics, ranging from simple to complex. For example, The Council of Supply Chain Management Professional defines logistics as the,"part of supply chain management that plans, implements, and controls the efficient forward and reverses flow and storage of goods, services and related information between the point of origin and the point of consumption in order to meet customer's requirements (Council of Supply Chain Management Professionals, 2011)"

It can be inferred from these definitions that logistics is getting a product from point A to B during the course of business operations. What times together this network of manufacturers, distributors, retailers, and transportation companies? The entire globe is connected through Logistics Transportation Systems. Logistics Transportation Systems are considered all water, air, and land assets and routes utilized in the movement of cargo or freight across the globe from points of supply to points of demand. They are the glue that holds the entire global market together. All of the assets and routes falling under this broad definition will be discussed in the following text. This includes all equipment, infrastructure, and links that are incorporated into logistics transportation systems.

I.I Key Terms

Actual Capacity – greatest output capacity available with the given resource configuration.

Aggregate Planning – internal production planning function performed by companies; transportation volumes directly result from these forecasts.

Best Practice –process standard

Capacity – maximum potential output from a given resource configuration

Certification of Suppliers – grading suppliers to ensure that entities within a supply chain adhere to standard set by buyers, quality, time schedule, and pricing needs.

Change Management – strategic decisions to enable organization to quickly adapt to a dynamic global environment

Cost to Benefit Analysis – detailed analysis of the benefits directly resulting from costs associate with a specific decision.

Cost Effectiveness – measure of effectiveness of key strategic decisions

Continuous Process/Product Improvement – process of continuous evaluation and analysis of current resource configuration and utilization

Supply Chain Management – systematic coordination of business functions throughout an entire supply chain enterprise for the purposes of developing and maintaining operations improvements

Figure I: Container Storage Yard

1.2 **Importance of Transportation Systems**

There has been a buzz about the massive increase in connectivity throughout the globe in recent years. Academics have acknowledged that the growth of the global economy is driven by technology and transportation. Remote regions that have historically played only a small role in the global supply chain are now able to produce goods and deliver them to market at affordable rates. For example, from 1997-2007 U.S. exports to East African Countries (Burundi, Kenya, Rwanda, Tanzania, and Uganda) grew at a steady rate of 9.1% annually and imports from the same East African Countries grew at a rate of 7.7% (Beningo, 2009). While these countries do represent a small portion of the U.S. International Trade it is indicative of a larger trend of globalization, where markets that have traditionally been too expensive to service are now considered viable new markets for expansion. This indicates that these countries have increasingly entered the world market and are now considered a trading partner with the United States. With more markets comes increased competition throughout the world. Increased competition on the global scale implies that companies will continue to lean heavily on the skills of logisticians to increase efficiency through strategic supply chain choices that ultimately result in a larger bottom line with high profit margins. Table 2 reflects the importance of transportation costs implications throughout every process with the supply chain. It should illustrate the importance of maintaining efficient and economical transportation systems.

Table 2: Transportation Cost Effect on Various Operating Cost

	Warehousing Costs	*Order Processing and Information Costs*	*Lot Quantity Costs*	*Inventory Carrying Costs*
High Transportation Costs	Lower; less product required in the warehouse to maintain the desired service level	Higher; these costs are function of order size. High transportation costs imply faster shipment which imply smaller order sizes	Higher; Function of Manufacturing volume. High transportation costs imply faster shipment which imply smaller manufacturing lots	Lower; fewer products are required in the warehouse to maintain the desired service level.
Low Transportation Costs	Higher; more product is required in the warehouse to maintain the desired service level	Lower; these costs function of order size. Lower transportation costs imply slower transportation costs which imply larger order sizes	Lower; Function of Manufacturing volume. High transportation costs imply slower shipment which imply larger manufacturing lots	Higher; more product is required in the warehouse to maintain the desired service level.

Transportation systems innovations and improvements have also allowed isolated countries to compete on the global market and develop economies of scale, resulting in offering cheap products to market. Economies of scale result of companies having access to more and varied markets. Access to more markets implies access to more consumers. More consumers mean that production facilities are able to produce more products. The concept of economies of scale implies that companies that utilize more capacity within their existing facilities are actually able to offer their products to market at a lower price due to the increase volume of production and lower variable costs per unit.   These lower costs are carried over into the transportation portion of the supply chain. If higher volumes of product are shipped to a variety of markets then the actual price per unit is decreased. Equate this to the cost of shipping Christmas presents to your family. If you ship only one present to your family in a box you pay a standard rate. But is you consolidate all of your presents into one box, then ship it to your family you actually pay less when considering the number of items shipped.

Supply chain design is crucial in every supply chain but in some cases the system is actually more important than the product. Take for example supply chains designed to deliver perishable goods to market. Companies that specialize in time sensitive products, such as fruits and vegetables, must design and implement a supply chain that allows them to deliver fresh goods to consumers near and far. How else could bananas from Honduras wind up in super markets across the entire southeastern United States year round?

1.3 Transportation Modes

This chapter will provide and in depth discussion on the five major modes of transportation. Each of these modes has their own advantages and disadvantages, characteristics, and pricing. Table 3 provides a brief comparison of the major modes of transportation utilized in global supply chains. The five major modes of transportation that will be further discussed in this chapter include:

Table 3: Characteristics of Transportation Modes

Transportation Mode Comparison					
	Truck	Rail	Air	Water	Pipeline
Speed	Moderate	Slow	Fast	Slow	Moderate
Availability	High	Moderate	Moderate	Low	Low
Consistency	High	Moderate	Moderate	Low	High
Loss & Damage	Low	High	Low	Moderate	Low
Flexibility	High	Low	Moderate	Low	Low

***1.3.1 Truck* –** This mode of transportation has seen the most growth over the last 50 years due to trade liberalization. However with the large growth has resulted in significant problems to include: petroleum fuel consumption and pricing issues, increased environmental impact awareness, traffic congestion, and increase in loss and liability resulting from increased traffic accidents (Notteboom et al. 1998). Shipment by truck averages a relatively low cost and is subject to many regulations due to their time on state governed highways. The size and shape of the loads are also highly regulated by governing bodies. Limitations placed on freight shipments by road coupled with technical and economic factors limit "potential to achieve economies of scale (Notteboom et al. 1998)." Governing bodies dictate the maximum allowable weight for truck shipments within their sovereign borders. Refer to Table 4 for detail of allowable gross vehicle weight by country. This means that given the current operating environment, truck transportation will only provide a limited amount of savings resulting from high volume shipments.

Table 4: Highway Weight Restriction by Country

Freight Weight Limits by Country		
US*	20-22 tons	
Canada*	17-26 tons	
Brazil*	31 tons	
Russia*	18 tons	
China**	40 tons	
India*	23 tons	
Europe**	40 tons	
40' Containers		

*Road Limitations provided by ZIM Integrated Shipping Services, LTD.
**Notteboom et al. 1998

Freight Shipment by truck does enjoy some advantages over other modes of transportation. First, the trucking industry is very competitive. This benefits companies utilizing trucking companies because high competition leads to lower prices. This competition is the result of a relatively low capital investment in vehicles compared to other modes of transportation (Notteboom et al. 1998). "Low capital costs also ensure that innovations and new technologies can diffuse quickly through the industry (Notteboom et al. 1998)." Another advantage is that truck shipments are also considered timely and consistent because of high vehicle speeds. Third, road transportation provides much flexibility for companies in route selection. There are a multitude of routes that companies can ship products within the supply chain and the routing choices can lead to competitive advantages for efficient companies.

1.3.2 Rail — Rail transportation is characterized by a high level of economic and territorial control since most rail companies are operating in a situation of monopoly, as in Europe, or oligopoly, as in North America (Notteboom et al. 1998)." Rail is the choice of companies that need to ship goods long distance that are generally not time sensitive. This mode of transportation is considered cheap, efficient, and environmentally friendly. ""Rail Transport is a 'green' system, in that its consumption of energy per unit load per km is lower than road modes (Notteboom et al. 1998)." With containerized unit trains, economies of scale are quickly realized with the declining marginal cost of each additional container added to each shipment. Freight must be carried on fixed routes and all transshipment of goods must be conducted at rail terminals. Transshipment of freight from rail cars represents that majority of time consumed during rail transit. "Operating a rail system involves using regular (scheduled), but rigid, services (Notteboom et al. 1998)." Rail transport is also subject to governmental regulation regarding the maximum allowable weight of freight shipments and height restrictions.

Rail Transportation has a number of limiting factors in its implementation. The terminal requirements for rail hubs significantly increase the operating costs. These operating cost are correlated to the central location of the rail hubs where, often, located within urban areas. These operating costs can also be measured in the rail capital investment and annual maintenance fees associated with construction and upkeep. Secondly, rail transportation is constrained by physical geography. Freight rail transportation "rarely tolerates more than a 1% (Notteboom et al. 1998)," gradient during transit. Gradients during transit have fuel consumption and time implications. Additionally, rail transportation has struggled to adopt a universal standard for the gauge of rail used during construction. This may seem a small detail but in actuality it is very detrimental to the integration of transcontinental rail lines that cross multiple borders.

Table 5: Mode Comparison

Transportation Mode Comparison					
	Truck	Rail	Air	Water	Pipeline
Operations Cost	Moderate	Low	High	Low	
Market Coverage	Pt to Pt	Terminal to Terminal	Terminal to Terminal	Terminal to Terminal	
Competition	Many	Few	Moderate	Few	
Traffic Type	All Types	Low to Mod Value, Mod to High Density	High value, Low Density	Low Value, High Density	
Length of Haul	Short - Long	Medium to Long	Long	Med-Long	
Capacity (tons)	10-25 tons	50-12,000	5-12 tons	1,000-6,000	

While these constraints may seem like major flaws in rail transportation there remain a number of advantages to this mode of transportation. The first advantage is that freight shipment by rail is more efficient than road transport. Table 5 provides a general comparison of all modes of transportation and their unique characteristics and ideal operating scenarios. Rail transport can also easily be combined with other modes of transportation for freight shipments. Methods, such as "piggy backing," are used to efficiently combine road and rail freight shipments for inland shipment. This is just one of the ways that rail transportation can be modified for each shipment.

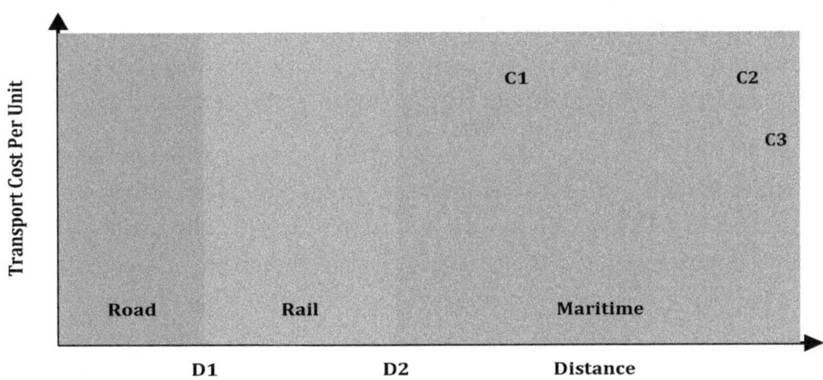

Figure 2:   Transportation Cost by Mode

1.3.3 Water – This mode of transportation is capable of hauling large quantities of product but is geographically limited. Water transport can be considered in two lights. The first is inland waterways that are somewhat limited by water and port access. Water transportation hinges on a number of set routes that have traditionally linked the world's ports together.

"Maritime routes are corridors of a few kilometers in width trying to avoid the discontinuities of land transport by linking ports, the main elements of the maritime/land interface. Maritime routes are a function of obligatory points of passage, which are strategic places, of physical constraints (coasts, winds, marine currents, depth, reefs, ice) and of political borders. As a result maritime routes draw arcs on the earth water surface as intercontinental maritime transportation tries to follow the great circle distance." (Notteboom et al. 1998)

The second is transcontinental shipments that consist of high volumes of product stacked on ship decks in TEU containers. Recently ships, known as Post Panamax, have been designed to ship as many as 13,000 TEU containers in one trip. This more than doubles the amount of cargo that can currently be shipped using the Panamax ship design.

I.3.4 Pipeline — Is high specialized and capital intensive. Capital investment depends on the size of pipeline and whether the pipeline is built above or below ground. This particular mode of transportation is usually reserved for oil and gas. Pipeline shipments are considered the most reliable of transportation modes because they have a dedicated supply and demand destination, such as an oil derrick to refinery. Pipelines have the ability to link isolated supply nodes with demand nodes where other means of transportation may not be economically viable. The major disadvantage of this mode of transportation is that it is by design inflexible. For example, a pipeline may be built to link a remote well that has a finite reserve of gas or oil. Once a particular supply node is exhausted, the dedicated pipeline then becomes obsolete.

I.3.5 Air - This particular mode of transportation is very expensive when compared to other modes. However, air freight is fast and reliable. Air freight is shipped in either the extra space on commercial airliners not occupied with passengers and their baggage or via dedicated air freight companies, such as UPS or Fed Ex. Since the mode of transportation is more expensive when compared with other modes it is usually reserved for time sensitive, perishable, or high priced items whose end consumer is a long distance away. Air freight has become more as the Just-In-Time inventory system has become more popular with all partners in supply chain operations, regardless of their positioning in the chain. "Air transportation's share of world trade in goods is only 2% when measured by weight but more that 40% when measured by value (Notteboom et al. 1998)." Table 6 provides compares the major modes of transportation from a speed and expense standpoint. As you can see air freight shipments are by far the most expensive mode of transportation. Air freight requires a large capital investment and operating budget in order to compete on the global scale. However, operating costs and capital investments are rewarded with high margins of revenue. For instance, "for international operations, freight can account to 45% of the revenue of a regular airline (Notteboom et al. 1998)." In the case of commercial airliners, freight shipments are an excellent method of recouping lost revenue resulting from variations in demand resulting from seasonality or other external factors.

Table 6: Cost Characteristics of Each Major Mode

Transportation Mode Comparison					
	Truck	Rail	Air	Water	Pipeline
BTU/Ton-Mile	2,800	670	42,000	680	490
Cents/Ton-Mile	7.5	1.4	21.9	0.3	0.27
Avg Length of Haul	300	500	1,000	1,000	300
Avg Speed (MPH)	40	20	400	10	5

Air freight shipments enjoy more freedom of movement when compared to other modes of transportation. They may go in the most direct path toward their final destination in any direction. However, air freight and commercial airliners are generally relegated to specific fly zones over the air space of specific countries. The majority of major carriers now utilize the Hub-and-Spoke method of supply chain design. This means that major companies operate a major hub, centrally located within their area of operations, which is connected to smaller regional hubs strategically placed within the same area. The Hub-and-Spoke distribution design allows companies to operate at the lowest cost per parcel shipped possible by concentrating operating costs in a main location.

Lecture 2

Power Point Handouts

Transportation Networks

Dr. MD Sarder

Transportation Network

➢is typically a network of transportation equipment and infrastructure which permits either vehicular movement or flow of goods. It consists of –

➢Nodes (facility locations or terminals) - Nodes are junction points of links.
➢Links (shipment flows or routes) - Links are the connections between the nodes of a network.

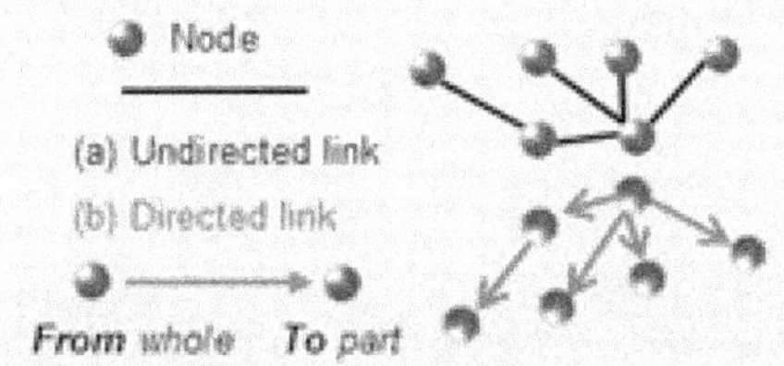

Layers of Networks

> **Physical Network:** The actual path that the product takes from origin to destination. Basis for all costs and distance calculations – typically only found once.

> **Operational Network:** The route the shipment takes in terms of decision points. Each arc is a specific mode with costs, distance, etc. Each node is a decision point.

> **Strategic Network:** A series of paths through the network from origin to destination. Each represents a complete option and has end to end cost, distance, and service characteristics.

Physical Network

- Guideway
    - Free (air, ocean, rivers)
    - Publicly built (roads)
    - Privately built (rails, pipelines)
- Terminals
    - Publicly built (ports, airports)
    - Privately built (trucking terminals, rail yards, private parts of ports and airports)
- Controls
    - Public (roads, air space, rivers)
    - Private (rail, pipelines)

Operational Network

◆ Four Primary Components
- Loading/Unloading
- Local-Routing (Vehicle Routing)
- Line-Haul
- Sorting

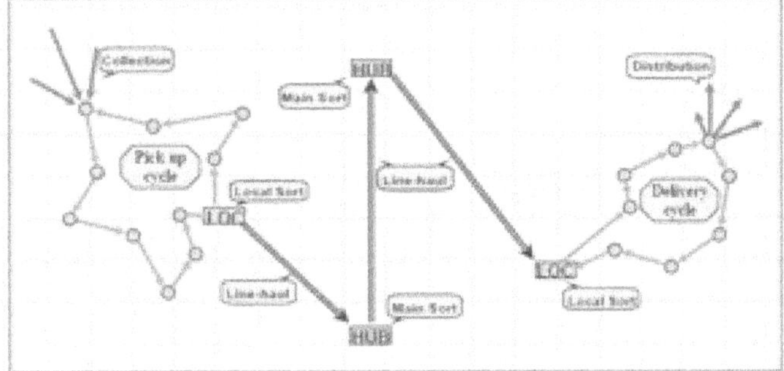

Node & Arc view of network
Each Node is a decision point

Strategic Network

◆ Path view of the Network
◆ Used in establishing overall service standards for logistics system
◆ Summarizes movement in common financial and performance terms – used for trade-offs

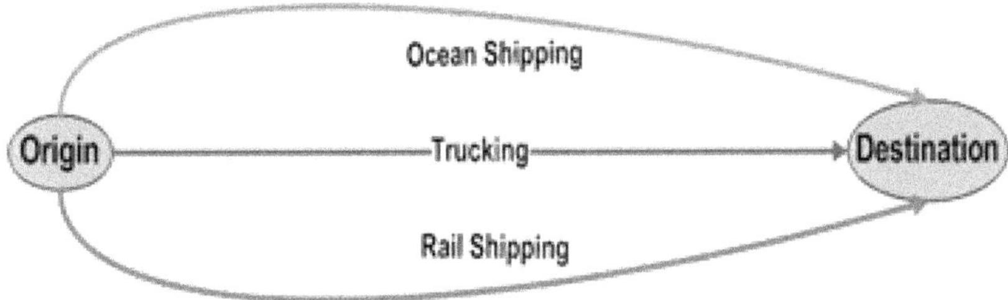

Primary Activities

➤ Loading/Unloading
➤ Line-Haul/ Back-Haul
➤ Local-Routing (Vehicle Routing)
➤ Sorting

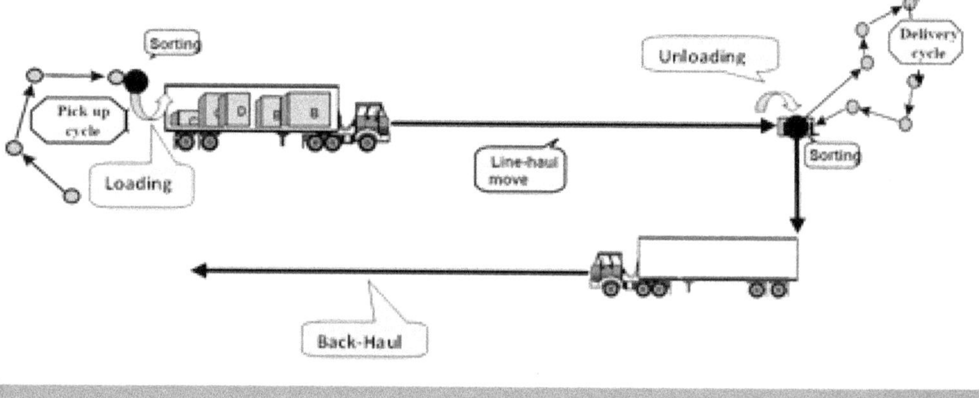

Primary Activities

◆ Loading/Unloading
- Key drivers:
  - Number of items
  - Time
  - Stowability (Packaging)
- Not always symmetric

◆ Linehaul
- Key drivers:
  - Distance
  - Balance / Backhaul
- Impacted by network
  - Congestion
  - Connectivity

Primary Activities

◆ **Vehicle Routing**
- Key drivers:
  - Number/Density of stops
  - Vehicle Capacity
  - Time
- Origin or Destination
  - One to Many
  - Many to One
  - Interleavened

◆ **Sorting**
- Key drivers:
  - Stowability (Packaging)
  - Number of items
  - Timing (Banking)

Transportation Cost Function

➢ Cost per item
  = Holding cost + Moving cost
  = Inventory cost + Shipment cost + Handling cost

➢ Holding/Inventory cost
  = Inventory holding cost X Avg. annual inventory
  = $rv[Q/2]$ = $rv[TD/2]$

➢ Inventory cost per item = $rv[T/2]$

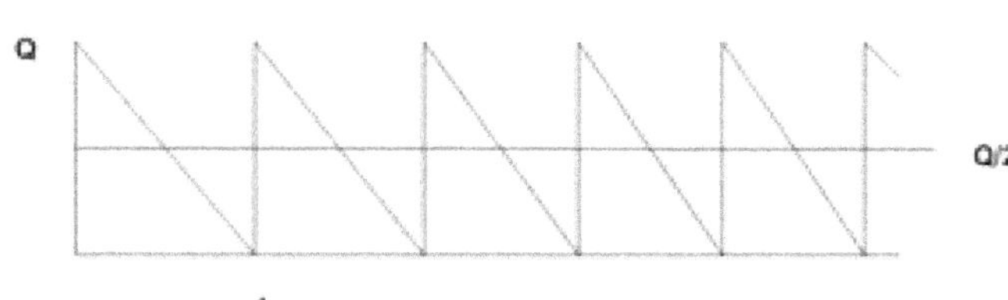

Transportation Cost Function

> Shipment/Transport or Movement cost

$$ShipmentCost = c_f + c_v Q$$

$$c_f = c_s(1+n_s) + c_d d \qquad c_v = c_{vs} + c_{vd} d$$

$$ShipmentCost = \left[c_s(1+n_s) + c_d d \right] + \left[Q(c_{vs} + c_{vd} d) \right]$$

$$TransportCPI = c_s\left(\frac{1+n_s}{Q}\right) + c_d\left(\frac{d}{Q}\right) + c_{vs}$$

Where
A = Fixed order cost (\$/shipment)
r = Inventory holding cost (\$/yr)
v = Purchase cost (\$/item)
Q = Shipment size (items)
Q_h = Handling size (items)
Q_{hmax} = Maximum handling size (items)
D = Annual demand (items)
T = Shipment frequency (yr) = Q/D
L = Lead time for transport (yr)
c_f = Fixed transport cost (\$/shipment)
c_v = Variable transport cost (#/item)
c_{fh} = Fixed handling cost (\$/shipment)
c_{vh} = Variable handling cost (#/item)
c_s = Fixed cost per stop (\$/stop)
c_d = Cost per distance (\$/distance)
d = Distance traveled
c_{vd} = Marginal cost / item / distance
c_{vs} = Marginal cost / item / stop
n_s = Number of delivery stops

Transportation Cost Function

> Handling cost

$$HandlingCost = c_{fh} + c_{vh} Q_h$$

> Shipment & handling cost per item

$$Transport\ \&\ Handling = c_s\left(\frac{1+n_s}{Q}\right) + c_d\left(\frac{d}{Q}\right) + \left[c_{vs} + c_{vh} + \frac{c_{fh}}{Q_{hMAX}} \right]$$

> Total cost per item

$$rv\left(\frac{T}{2}\right) + c_s\left(\frac{1+n_s}{Q}\right) + c_d\left(\frac{d}{Q}\right) + \left[c_{vs} + c_{vh} + \frac{c_{fh}}{Q_{hMAX}} \right]$$

Sample Problem

➤Dell Computer Inc. bought $525 of materials from its vendor for a particular order of 5 computers. To deliver this particular order, Dell driver needs to stop in 4 different places to cover a total distance of 250 miles, the cost per stop is $0.43 and cost per mile is $0.5 since it is using UPS service for delivery. Marginal cost per stop is $0.25, fixed cost of handling a pallet of computers (pallet size = 5) is $10 and variable cost of handling the same pallet is $7.50.

➤How much will Dell charge its customer for shipping and handling for each computer?
➤What would be the total cost per computer if Dell holds its inventory on an average 2 weeks at $0.25/year against every $1 of inventory?

Sample Problem

➤Following information is given;

r = Inventory holding cost ($/yr) = $0.25/yr = [$0.25/52] X 2 = $0.0096
v = Purchase cost ($/item) = $525/5 = $105
Q = Shipment size (items) = 5
Q_h = Handling size (items) = 5
Q_{hmax} = Maximum handling size (items) = 5
c_{fh} = Fixed handling cost ($/shipment) = $10/shipment
c_{vh} = Variable handling cost (#/item) = $7.50/5 = $1.5/computer
c_s = Fixed cost per stop ($/stop) = $0.43/stop
c_d = Cost per distance ($/distance) = $0. 5/mile
d = Distance traveled = 250 miles
c_{vs} = Marginal cost / item / stop = $0.25/5 = $0.05/computer/stop
n_s = Number of delivery stops = 4

Sample Problem

1. Total shipment & handling cost per computer

$$Transport\ \&\ Handling = c_s\left(\frac{1+n_s}{Q}\right) + c_d\left(\frac{d}{Q}\right) + \left[c_{vr} + c_{vh} + \frac{c_{fh}}{Q_{hMAX}}\right]$$

$$= \$0.43\left(\frac{1+4}{5}\right) + \$0.5\left(\frac{250}{5}\right) + \left[\$0.05 + \$1.5 + \frac{\$10}{5}\right]$$

$$= \$28.98$$

Given;

$r = \$0.0096$
$v = \$105$
$Q = 5$
$Q_h = 5$
$Q_{hmax} = 5$
$c_{fh} = \$10$
$c_{vh} = \$1.5$
$c_s = \$0.43$
$c_d = \$0.05$
$d = 250$
$c_{vs} = \$0.05$
$n_s = 4$

2. Total cost per computer
   = Inventory cost + Transport & Handling cost
   = $rv[Q/2]$ + \$28.98
   = \$0.0096 X 105 X 5/2 + \$28.98
   = \$31.50

Influence on Shipment Cost

1. Size or volume of shipment

◆ For an individual shipment –
  - Captures allocation of fixed costs over many items
  - Follows lot sizing logic – drives mode selection

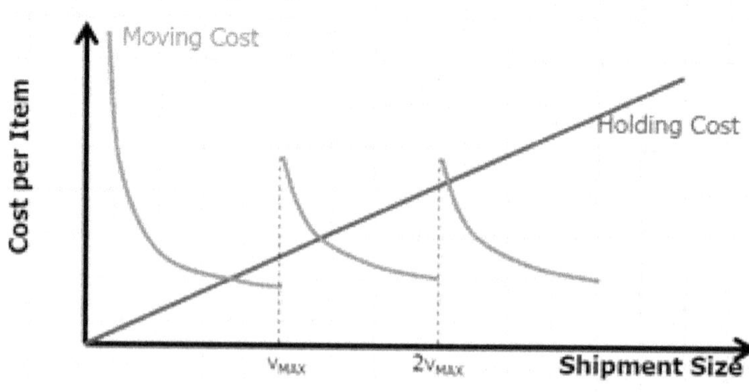

Influence on Shipment Cost

2. Load or flow balance

◈ Reverse flow mitigates the cost of repositioning.
◈ Strong for direct carriers – but present in all
  - Subadditivity - the costs of serving a set of lanes by a single carrier is lower than the costs of serving it by a group of carriers
  - Cost Complementarity - the effect that an additional unit carried on one lane has on other lanes

Influence on Shipment Cost

3. Location and shipment density

◈ Strong for Consolidated Carriers
  - Location Density
    - Number of customers per unit area
  - Shipment Density
    - Average number of shipments at a customer location
    - Daily average volume is critical

Which is better?

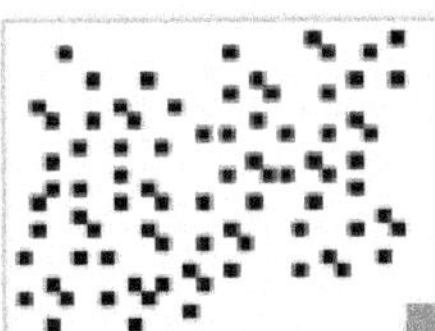

Lecture 2
Reading Materials

2.0 Introduction

"A network is distinguished from a pattern or grid, a weave or a series of overlays in that it connects things to achieve something...In transportation, the strands are routes and the knots are places and the goal is moving people, goods and services as efficiently and cost-effectively as possible to increase prosperity and opportunity" (CNU, 2008, p. 3). Transportation networks play a vital role in any society – the movement of good and people are essential. The analysis of transportation networks is important because "mobility and accessibility are major determinants of lifestyle and prosperity" (Bell & Iida, 1997, p.I). Transportation networks include a system of roads, streets, pipes, aqueducts, power lines, rail lines, or other structures that allow movement of commodities. Transportation network analysis is employed to determine the flow of commodity through the network and can be analyzed using various methods.

2.1 Terminology

In order to discuss transportation networks, it is important to first define some terminology and types of networks. *Pure networks* include a network that is only concerned with topology and connectivity. In a *flow network*, in addition to topology and connectivity, flow properties are considered. Flow properties include origin-destination demands, capacity constraints, path choice and costs of creating links. "A transportation network is a flow network representing the movement of people, vehicle or goods" (Bell & Iida, 1997, p.3). The transportation network is made up of *links* (movement between locations including time) and nodes.

Links may be *directed* (have a specified direction of movement) or *undirected*. *Link length* describes the length between nodes or in vehicle average. *Link cost* is typically used to describe the "linear combination of time and distance" and *link capacity* is the maximum flow of commodities for the link (Bell & Iida, 1997, p. 17). *Movement* is defined as the movement of commodities from a distinct origin and destination (house, office, zone, etc.). *Centroids* are either origin or destination nodes or *internal nodes* are things like distribution centers.

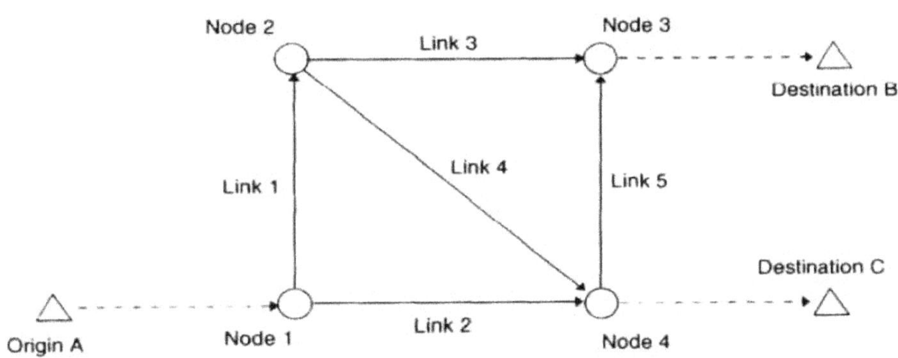

Figure I: Example of a transportation network (Bell & Iida, 1997, p. 18)

Figure I from Bell and Ida (1997) is an example of a transportation network that contains one origin centroid (Origin A), two destination centroids (Destination B and C), five links (Node I to 2; Node I to 4; Node 2 to 4; Node 2 to 3; Node 4 to 3), four internal nodes (I, 2, 3, 4), three connectors (Origin A to Node I; Node 4 to Destination C; Node 3 to Destination B) and 5 paths (Node I to Node 4 to Node 3 to Destination B; Node I to Node 2 to Node 4 to Node 3 to Destination B; Node I to Node 2 to Node 3 to Destination B; Node I to Node 4 to Destination C; Node I to Node 2 to Node 4 to Destination C)

. The links are all directed in Figure I (specified direction of movement). In addition to paths, there are *cycles* in which a path connects to itself at the ends, *trees* in which each and every node is visited once and only once, and *cutsets* that describe minimal links that when removed from the network would divide the network in two resulting in an absence of links between the two resulting sub-networks (Bell & Iida, 1997). Network transportation terms are clarified by Bell and Iida (1997, p. 19) in Table I.

Table I by Bell & Iida (1997) definition of terms.

Concept	Definition
Node	Junction of two or more links. It is either an *internal node* (neither source nor sink) or a *centroid* (either source or sink)
Link	Conduit for flow between two nodes
Connector	Link between a centroid and an internal node
Movement	Flow with a specified origin and destination
Path	A sequence of nodes connected by links in one direction so that a movement is feasible from the first node to the last node in the sequence. Often the first and last node in the sequence are centroids
Cycle	A path with the same node at either end
Tree	A network where each node can be visited once and only once
Cutset	A set of links whose removal would lead to two sub-networks with no connections
Commodity	A kind of flow distinguished by origin, destination or some other feature
User class	A kind of flow distinguished by its behaviour, for example its sensitivity to cost

In addition to these terms, it is important to clarify some of the different types of transportation networks. In *linear networks*, there are no path choices but there origins and destinations (i.e., a railway line). In *a grid network*, there are multiple origins and destinations with varying route choices (i.e., urban area). Networks get increasingly complex as paths, nodes, modes—individual modes (like cars or bicycles) and communal modes (like buses)—turning movements and junctions are added. Today's transportation network analysis is heavily dependent on computer software and algorithms.

Wilson and Nuzzolo (2008) state that the "the main efforts to reduce solution time (especially for assignment problems) are in improving algorithms and procedures for path search in space-time networks" (p13). Some models when coupled with algorithm—like Nielson's model and Dijkstra's algorithm—lead to gains but is based on a heuristic solution. Friedrich's algorithm is another example of techniques used to conduct transportation analysis. Some examples of commercial software solutions include EMME/2, VISUM, and OMNITRANS and some examples of non-commercial products include SASM, SAVEF, and TPSCHEDULE (Wilson and Nuzzolo, 2008). The following is a summary of selected transportation network analysis models and algorithms.

2.2 Algorithms

Algorithms are viewed as a way to efficiently optimize transportation problems with computers. "The basic concept of a routing algorithm is to model the specific problem in a suitable graph ant to compute a shortest path to solve it" (Geisberger, 2011, p. 11). The more complex a transportation problem becomes the harder it is to identify or create an effective algorithm.

One of the most widely known algorithms is Dijkstra's Algorithm, a shortest path algorithm that finds the shortest path from a "single source node to all other reachable nodes in the graph by maintaining tentative distances for each node" (Geisberger, 2011, p. 13). The algorithm puts the nodes in order of the shortest distances.

Another method is ALT—A_, Landmarks and the Triangle inequality (ALT). ALT is an algorithm that uses landmarks to compute shortest-path distances. It can be combined with other algorithms to increase speed (Geisberger, 2011). Edge labels involve pre-computing information for an edge that specifies a group of nodes when that group of nodes is a superset of all nodes on the shortest path. This algorithm has been developed into more detailed section graphing (Geisberger, 2011). The heuristic algorithm proposed by Zografos and Androutsopoulos (2008) is designed to address itinerary planning problems—wait time, best route, travel time, etc.—and can be used by a web-based application for journey planning.

One nonlinear optimization algorithm is the Frank and Wolfe method in which the solution is created by minimizing the "original objective function over the line segment connecting the current solution and the sub problem solution" (Bar-Gera, 1999, p. 8). This type of algorithm is most commonly used in traffic assignment problems. Related link-based algorithms have also been used to address traffic assignment problems. One example of this is Hearn's Restricted Simplicial Decomposition method that uses a multi-dimensional search over the convex hull of the earlier sub problem solutions (Bar-Gera, 1999, p. 9). Route-based methods and origin-based algorithms also have been developed and employed.

2.3 Mathematical Tools

Possibly the most appropriate and useful field of math for transportation problems is graph theory. It allows us to model locations and routes between those locations. Costs can be attached to the links in a network to account for fuel cost, travel time, or anything else relevant to the problem.

Standard algorithms such as Dijkstra's can be used to efficiently find the least costly routes between nodes in a graph. By formulating real-world problems into a mathematical form, we can use this and other methods to discover properties of the network that might not be obvious. For instance, some underutilized back road might prove to be part of a less costly route.

For many transport problems, linear programming methods can be used to find optimal solutions. When more than one production point can supply more than one customer, LP helps us to determine how much each producer should send to each customer endpoint. Production capacities and customer needs are given as linear constraints. Transportation costs (using same underlying data as graph theory approaches) are used as coefficients of an objective function used to minimize overall cost.

2.4 Transportation Modes and Methods

Transportation deals with the movement of goods from a source such as a plant or factory to a destination such as a warehouse or store. Transportation has many modes of operation using airways, waterways, railways, roadways. Planes, boats, trucks, and trains are the vehicles used in transportation. The goal of suppliers is to minimize transportation costs while meeting customer demand for a product. Generally, transportation costs depend upon the distance traveled between the source and the destination, the modes of transport chosen, and the size and quantity of the product to be shipped. Many combinations of these variables can affect transportation cost and must be looked upon carefully.

2.5 Exploring Modes of Transportation

Modes	Pros	Cons
Road	• Flexible route options • Cost effective • Ideal for short distances • Convenient	• Subject to traffic delays • Incapable of overseas shipment • Somewhat fixed load capacity
Rail	• Fast delivery • High load capacity • Cost effective	• Inflexible route options • Limited routes and timetables • Transport to and from depot can add cost
Water	• Easily handle heavy loads and large capacity • International shipment capability	• Long lead time • Limited ports • Inflexible routes • Transport to and from ports can add cost
Air	• Fast delivery • International shipment capability	• Very costly • Limited routes

- Road – major role in transportation, trucks used, trucks have different load capacities
- Railways – Large cargo moved over short period of time, many stops, few endpoints
- Waterways – cheapest means of transport overall, international trade done through ships, carry large volumes, slow
- Air – fastest mean of transport, most costly

Travel distance, destination, route selection and lead time must be factored in to the final decision when choosing the best mode of transportation for a particular shipping need. Multiple modes occasionally must be called upon to fulfill the shipping needs, especially when using rail or water options due to their capability of only reaching depots or ports and not warehouses or stores.

The routes these modes of transportation used can be thought of as networks. Along these networks are *nodes*. Nodes are, effectively, stops – or connections - in a transport system. Transportation networks connect these nodes. The networks chosen have an effect on the cost and efficiency of the transportation of goods. These networks can be viewed from three angles. Each angle may have unique effects on shipment decisions.

2.6 Network Elements

- Physical network
- Operational network
- Strategic network

The *physical network* is the physical path the shipment takes. The *operational network* is like an abstraction of the underlying physical routes. Only decision points are represented, not every road intersection. And, finally, the *strategic network* is the grand scale among the three – looking at whole routes as individual links with total costs attached.

2.7 Network Problems

Bottlenecks often exist in a transportation network. These are areas of the network that are essential for getting shipments from one side to the other. Since these areas must be used, there could be contention among various parties that need to share routes. Transportation modes such as rail have more centralized scheduling to meet the needs of users. On the other end of the spectrum, public roadways are dependent on collective behavior of not only other shippers but a mass of passenger traffic as well. This makes scheduling for some modes less predictable than others.

Bottlenecks can occur either on links (busy roadways, bridges) or nodes of a network (unloading/sorting/reloading at cross docking facilities for example). Some are dictated by local geography: bridges over rivers, narrow passages between mountains, tunnels, and so on. Some are temporary, like road repair work on a busy highway.

How do we identify bottlenecks in order to plan ahead? Graph theory gives us the notion of "cut points", which are links that if removed would make a route from one area to another impossible. Essentially, a bottleneck is a heavily used cut point. Reducing the impact of bottlenecks through construction projects is more of a city planning or civil engineering problem, and we concern ourselves simply with using existing infrastructure as it is.

Conclusion

Transportation networks are a crucial part of the commercial industry. Network modeling methods are an integral part of planning the best use of transportation systems. These models are used to translate a complex problem into a scenario that can be used to calculate and manipulate costs and scheduling for delivery of goods.

Transportation networks must be analyzed and modeled to meet an individual company's needs based on factors such as the product they ship and the location to which they ship. There is no universal model to best fit with ever-changing variables.

References:

1. Bar-Gera (1999). Origin-based algorithms for transportation network modeling. Technical Report Number 103 for NISS.
    http://www.niss.org/sites/default/files/pdfs/technicalreports/tr103.pdf
2. Bell, M. H., & Iida, Y. (1997). *Transportation network analysis [electronic resource] / Michael G.H. Bell, Yasunori Iida.* Chichester; New York: J. Wiley, c1997.
3. CNU (2008). Defining and Measuring Sustainable Transportation Network. CNU Transportation Summit; Charlotte, NC November 6-8, 2008.
    http://www.cnu.org/sites/files/defining_measuring_sustaining.pdf
4. Geisberger, R. (2011) Advanced Route Planning in Transportation Networks (dissertation).
    http://algo2.iti.kit.edu/download/diss_geisberger.pdf
5. Wilson, N. & Nuzzolo, A. (2008). Schedule-Based Modeling of Transportation Networks. Dordrecht: Springer.
6. Zografos, K. & Androutsoulous, K. (2008). Algorithms for Itinerary Planning in Multimodal Transportation Networks. *Intelligent Transportation Systems, 9*(1).

LECTURE 3

POWER POINT HANDOUTS

Transportation Networks
Direct vs. Hub

DR. MD SARDER

Direct Network

> Shipping from origin to destination without transshipment

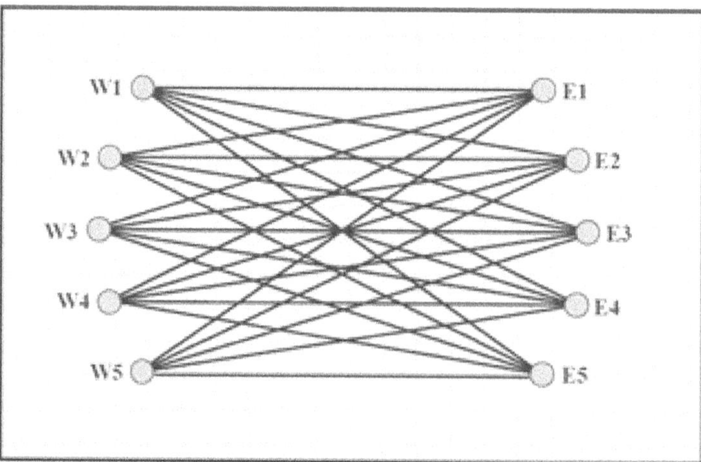

Direct Vs. Hub Network

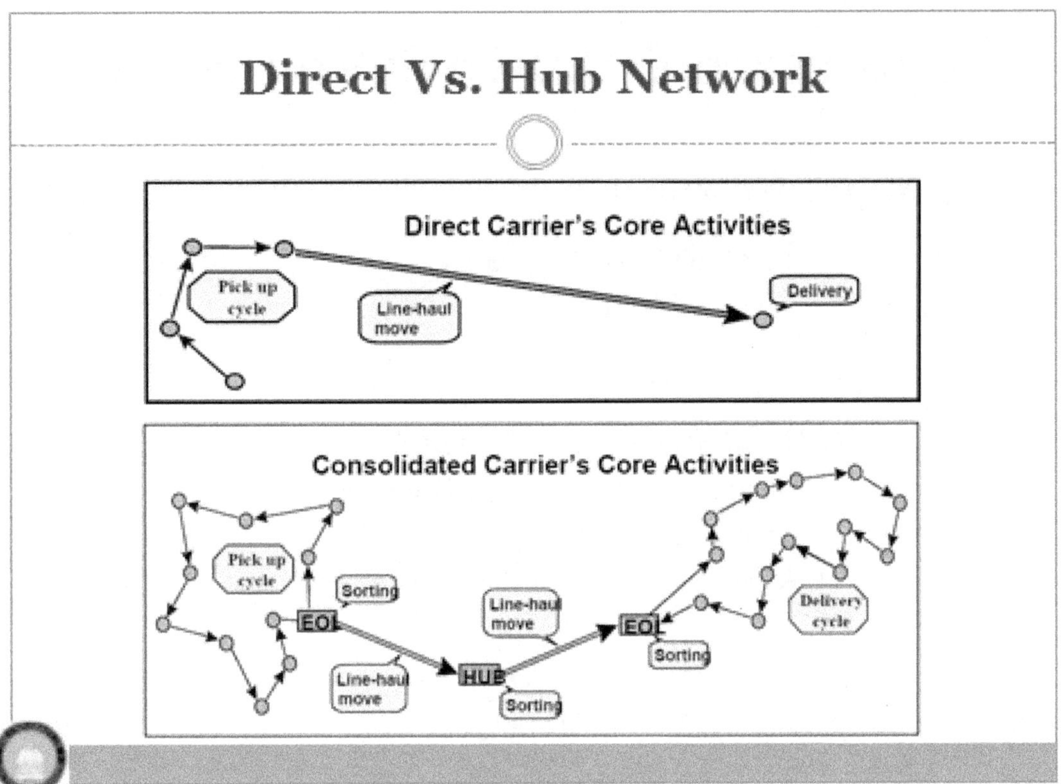

Operational Network Structure

◆ One to One
- ■ Direct Network

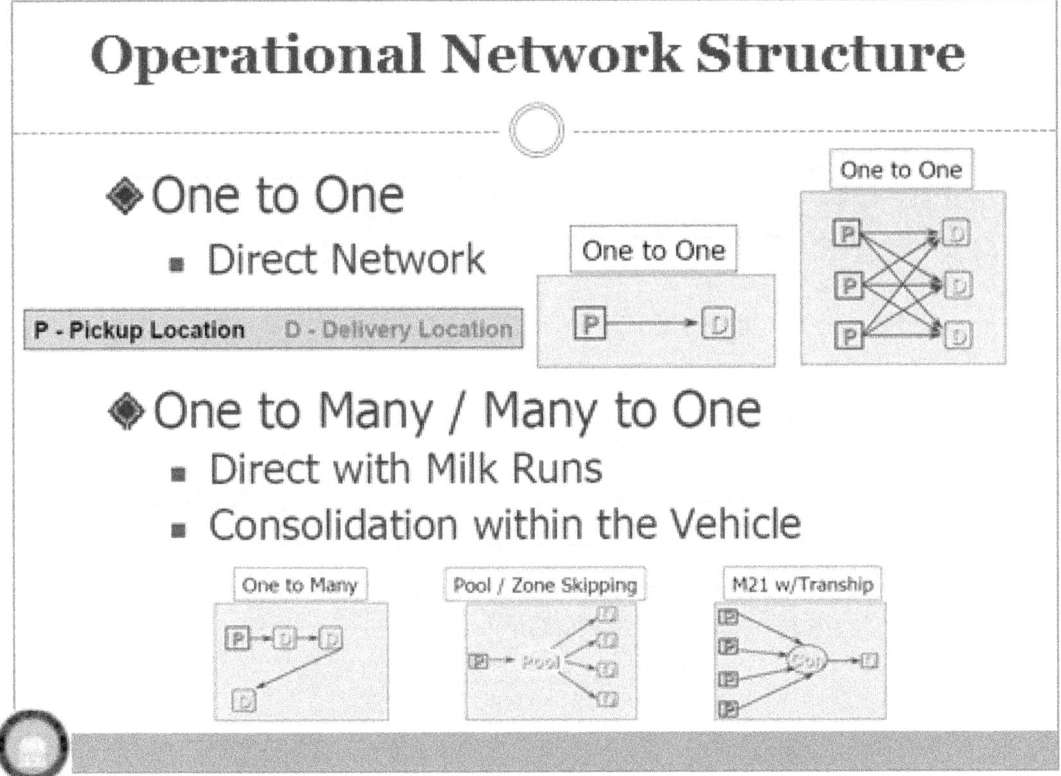

◆ One to Many / Many to One
- ■ Direct with Milk Runs
- ■ Consolidation within the Vehicle

Operational Network Structure

◆ **Many to Many**
- **No Transhipment Point**
  - Direct with Milk Runs

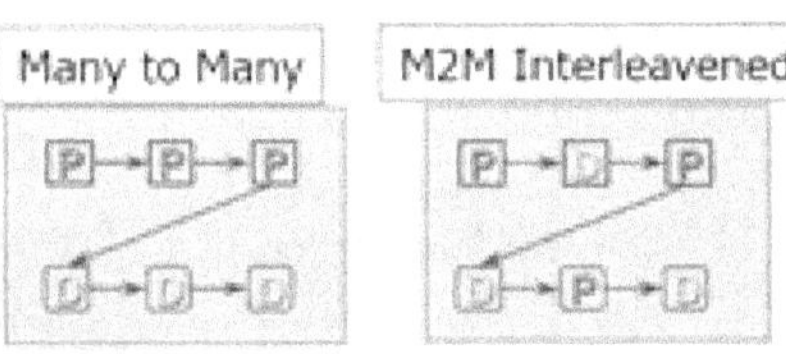

| P - Pickup Location | D - Delivery Location |

- **With Transhipment Point**
  - Direct with DC (Cross Docking)
  - Direct with Milk Runs

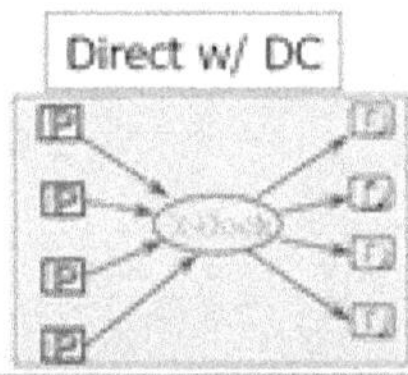

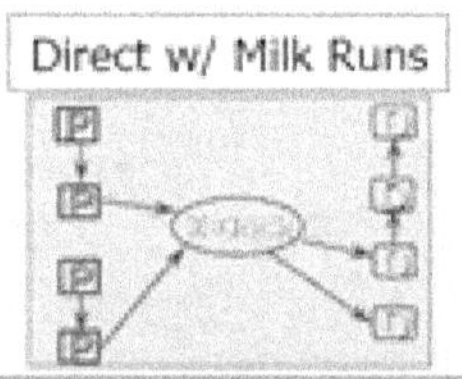

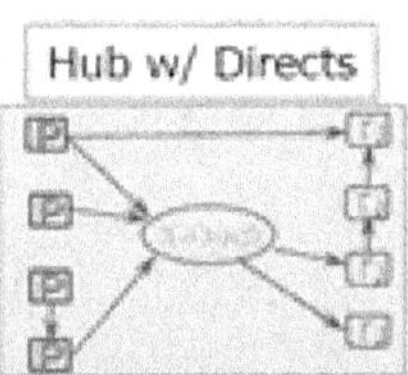

Performance Criteria

➢ Number of trucks
➢ Number of trips
➢ Shipment cost
➢ Frequency of service

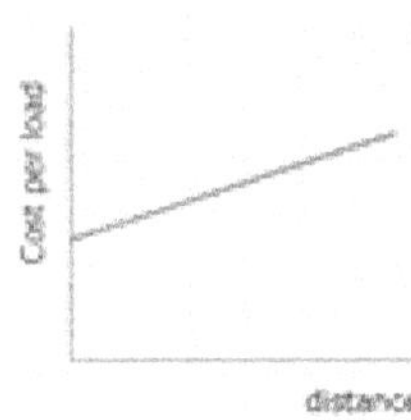

➢ Example Case
- ✓ Pick up & delivery every day from terminals to destination
- ✓ Total demand for 5 customers = 2TL delivery
- ✓ Average distance between terminals = 500 miles
- ✓ Average distance from terminals to hub = 250 miles
- ✓ Cost for transportation = $200 handling + 1 $/mile
- ✓ Cost for using hub = $100/day

Performance Criteria

> Example Case: Direct Shipment
> - Total no. of shipments = 5
> - Total handling cost = 5 X \$200 = \$1000
> - Total shipment cost = 5 X (\$1 X 500) = \$2500
> - Total cost = \$1000 + \$2500 = \$3500
> - Level/frequency of customer service?

> Example Case: Via Hub Shipment
> - Total no. of shipments = 2 + 5 = 7
> - Total handling cost = 7 X \$200 = \$1400
> - Total shipment cost = 7 X (\$1 X 250) = \$1750
> - Hub cost = \$100
> - Total cost = \$1400 + \$1750 + \$100 = \$3250
> - Level/frequency of customer service?

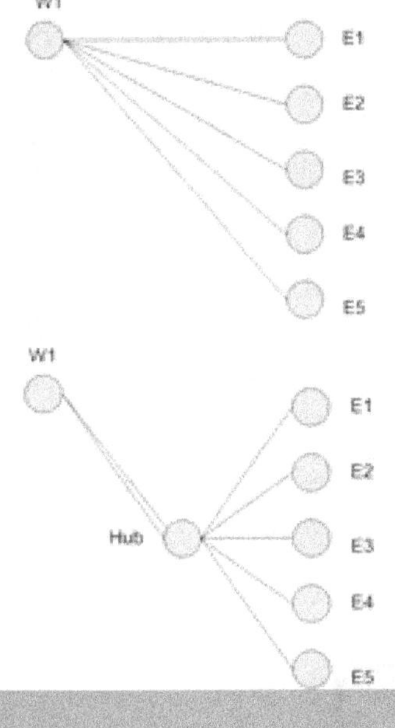

Hub Advantages

- Hub consolidation reduces costs
  - Consolidation increases conveyance utilization
  - Transportation has a fixed (per conveyance) cost
- Fewer conveyances are required
  - Is consolidation better . . .
    - when point to point demand is higher or lower?
    - when variability of point to point demand is higher or lower?
  - Coefficient of variation as useful metric
- Provides better level of service with fewer resources
  - Non-stop vs. frequency of service
  - Non-stop vs. geographical coverage
    - serving more / smaller cities

Hub Advantages

- ◆ Relative distances
  - Degree of circuity
- ◆ Vehicle and shipment size
  - Smaller shipments → hub more economical
- ◆ Demand pattern
  - Many destinations from each origin
  - Many origins into each destination
- ◆ The hub location
  - Significant business generation for passengers
    - Air – large city
    - Transit – CBD
  - Good access for freight
    - Highways access
    - Away from population centers

Hub Disadvantages

- ◆ Cost of operating the hub
  - Facility costs
  - Handling costs - unloading, sorting, loading
  - Opportunity for misrouting, damage, theft (shrinkage)
- ◆ Circuity
  - Longer total distance travelled
  - More vehicle-hours expended
- ◆ Impact on service levels
  - Added time in-transit
  - Lower reliability of transit
- ◆ Productivity/utilization loss
  - Cycle/"bank" size

Bypassing Operations

◆ Considerations in setting direct service:

- Demand between E1 and W2
- Service E1-Hub and Hub-W2
- Effect on the hub
- Effect on E1 activities

> Heavy load
> Reliability & low cost
> Time constraints
> Pre packed or block load
> Cross docking

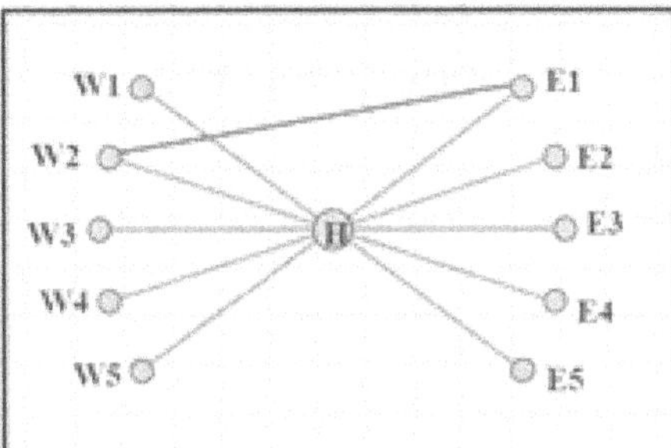

Regional Terminals

> Consolidation between regional terminals
> Regional hub
> Local routing
> Additional layer

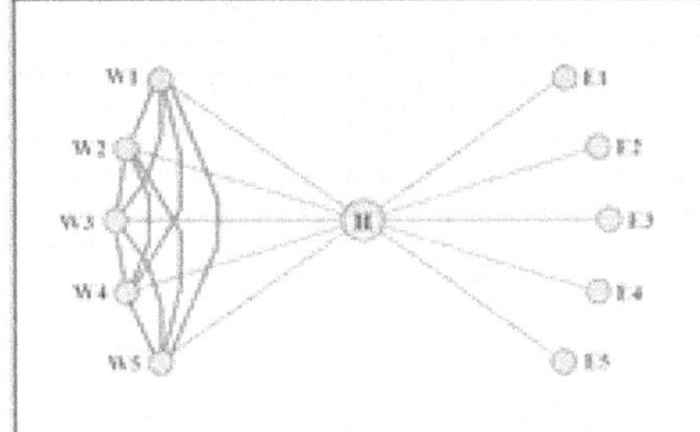

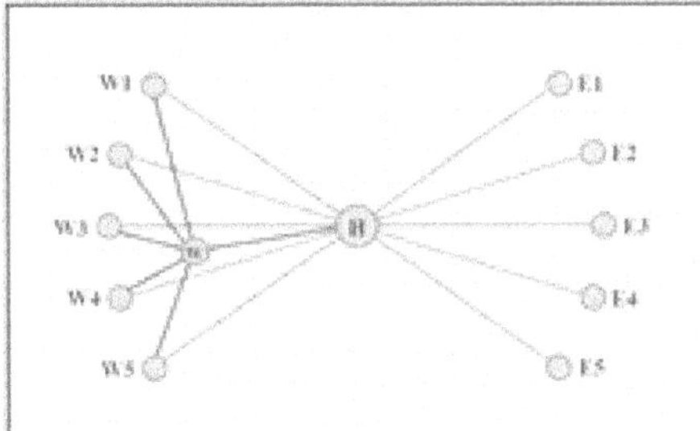

Hub & Terminal Bypassing

- More options
- Better responses
- More distance
- Under utilization of hub

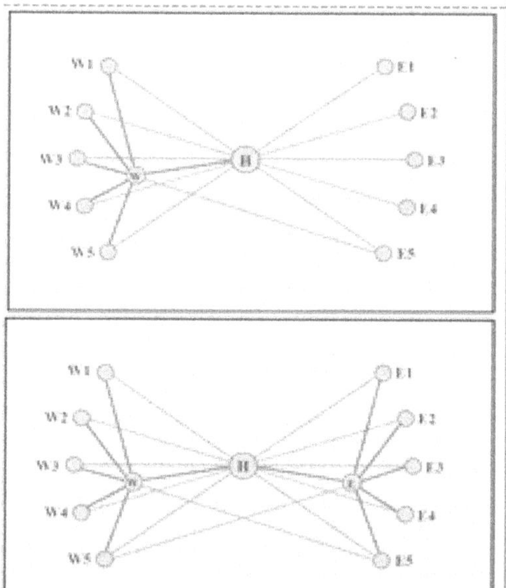

Routing Alternatives

Direct effects:
- On each of the three alternatives

Indirect effects:
- Congestion and spill-overs

Strategic Decision Making

◆ Service Offerings from W5 to E5
- Central Hub Routing
- Regional Terminal Routing
- Direct Routing

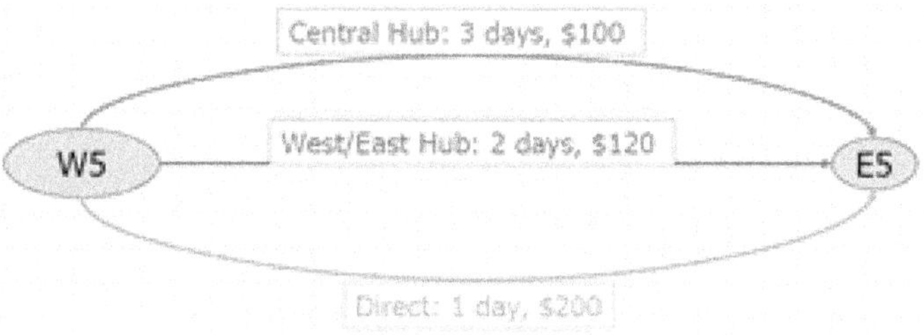

Decision Making: Tradeoffs

Structure	Pros	Cons
Direct Shipping	◆No intermediate DCs ◆Simple to coordinate	◆Large lot sizes (high inventory levels) ◆Large receiving expense
Direct w/ Milk Runs	◆Lower transport costs for smaller shipments ◆Lower inventory levels	◆Increased coordination complexity
Direct w/Central DC (holding inventory)	◆Lower IB transport costs (consolidation)	◆Increased inventory costs ◆Increased handling at DC
Direct w/ Central DC (X-dock)	◆Very low inventory requirements ◆Lower IB transport costs (consolidation)	◆Increased coordination complexity
DC w/ Milk Runs	◆Lower OB transport costs for smaller shipments	◆Further increase in complexity
Hybrid System	◆Best fit of structure for business ◆Customized for product, customer mix	◆Exceptionally high level of complexity for planning and execution

LECTURE 3
READING MATERIALS

Following topics have been covered in this section.

- Transportation network
- Direct network
- Hub and spoke network
- Direct versus hub and spoke
- Operational network structure
- One to one
- One to many/ Many to one
- Many to many – with transshipment point, without transshipment point
- Performance criteria
- Hub Advantages
- Hub Disadvantages
- Problems

3.0 Transportation Network

The term "network" is commonly used to describe a structure that can be either physical or conceptual. Each of these networks includes two types of elements: a set of points and a set of line segments connecting these points.

Transportation network is defines as the network of the infrastructure that includes the network of roads, pipes, aqueducts, power lines or any structure that permits vehicular movement or flow of the commodity. Transportation network analysis is used to determine the flow of vehicles that involves different modes of transportation. Networks consist of liner services and interconnection points, where the loading units are transferred between liner services. By linking liner services together to form a network, volumes can be attracted from a larger number of regions. Higher volumes enable operators to offer reasonable service frequency to and from areas with only modest transport demand. By consolidating cargo from different regions at a single or just a few points on the network, operators can optimize the deployment of their transport assets: they can decide on the frequency of service and on transport capacity for each of the links.

A disadvantage of networks is the higher total production costs of connecting origins and destinations. Transport via interconnection points in a network is synonymous with longer trip times and higher distances than with liner concepts. This raises the cost of labor, the capital costs of assets and distance-related costs, such as energy and maintenance. There are also third-party transshipment costs involved each time a loading unit has to transfer between liner services. Networks are based on the principle of a sequence of services connecting origins and destinations, which makes them rnore vulnerable to disruption

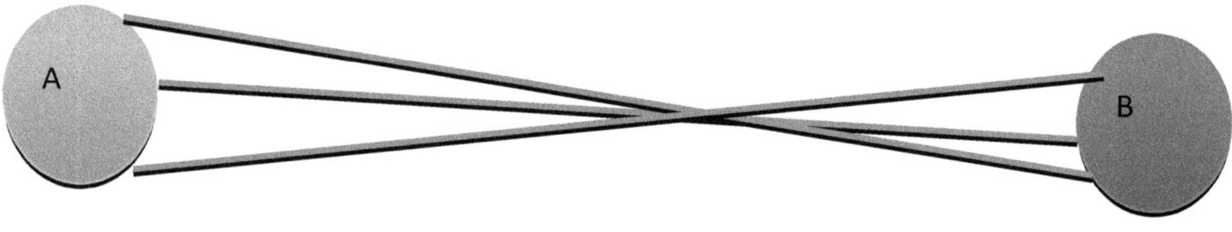

Figure I: Transportation Network

3.1 Classification of Transportation Network

The transportation network is classified as direct network and Hub and spoke network. These networks are further classifies as one to one, one to many, many to many networks with or without transshipment points. The various operational networks are designed and worked to increase the profit margins of the transportation systems.

Direct network is defined as the shipment of goods from the point of origin to the point of destination without transshipment. All the shipments come directly from suppliers to buyer's location. The advantages of direct network are they don't need intermediate warehouse and it is simple to coordinate. However, direct network involves higher inventories and significant receiving expenses.

The hub-and-spoke system is the best-known network system. The spokes in the network are liner services between regional terminals and the hubs. At the hub the transport units are transferred from one liner service to another connecting the hub with the destination terminal. Ideally, hubs are located near to the center of gravity of transport demand. In this way detour distances and trip times between origin and destination terminals can be minimized. The total terminal-to-terminal trip time is increased because of the extra distance for the call at the hub and the time spent in the hub itself. A hub-and-spoke system is designed to combine small flows arriving and departing in different directions.

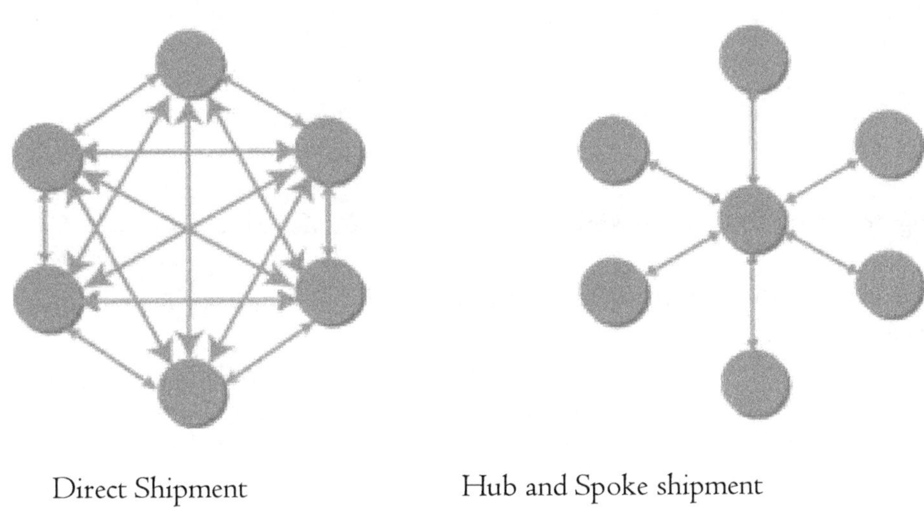

Figure 2: Direct Shipment and Hub & Spoke Shipment

The direct carriers core activities are pick up cycle, line haul and delivery of the shipments. The consolidated carriers core activities include consolidation of the shipments in the pick-up cycle, sorting of the products, line haul followed by hub consolidation, product sorting and delivery of the shipments.

3.3 Operational Network Structure

3.3.1 Direct shipping with milk runs

In this case, a supplier delivers directly to multiple buyer locations on a truck or a truck picks up deliveries destined for the same buyer location from many suppliers. This allows reduction in cost by eliminating the need for direct small shipments using LTL shipments. When deciding this option, a supply chain manager has to decide on the routing of each milk run.

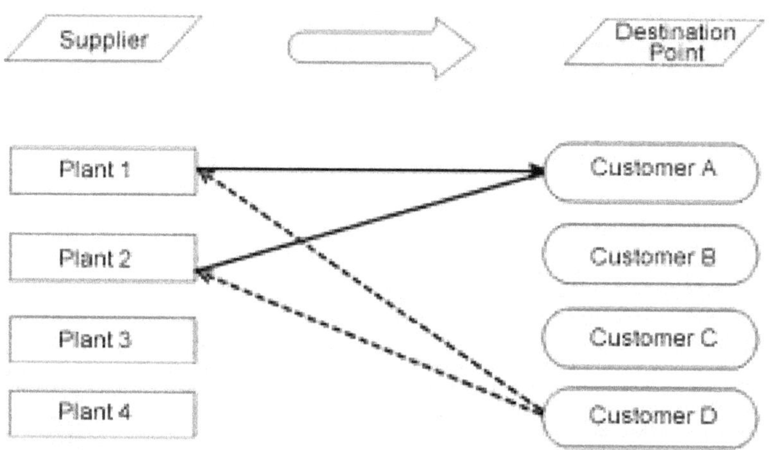

Figure 3: One to One Network (Source: Chopra & Meindl, 2004)

One to one shipment involves the shipping of goods from the point of origin to the point of destination that indicates

One-to-many shipment transportation will involve the movement of goods directly from the point of origin to the point of destination with or without transshipment points. It might also involve pool or zone skipping. Direct shipments with milk runs will lead to lower transport cost for smaller shipments and lower inventory levels. The disadvantages associated with direct shipping with milk runs causes increased coordination complexity.

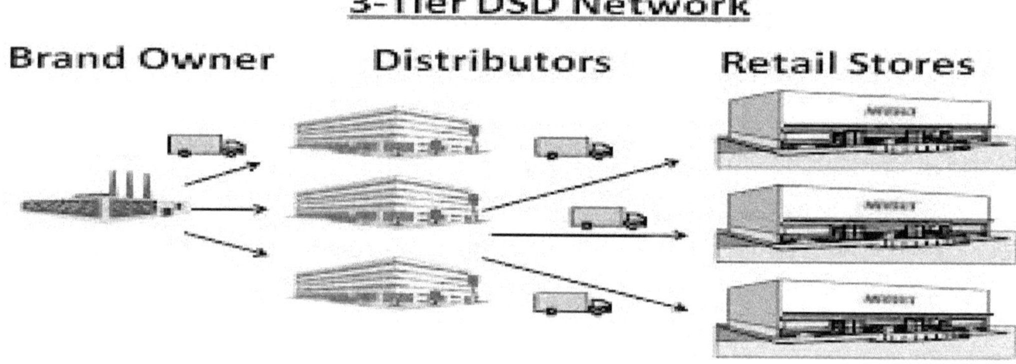

Figure 4: One to Many Shipments Network

Many to many transportation involves the multiple shipments of goods from different point of origins to the point of destinations. It may or may not include transshipment point. The Direct network with milk runs may or may not involve the transshipment point. The Cross docking distribution involves transshipment point.

3.3.2 All shipments via central DC

Under this option, suppliers do not send shipments directly to buyer locations. The buyer divides locations by geographic region and a DC is built for each region. Suppliers send their shipments to the DC and the DC then forwards appropriate shipments to each buyer location. A cross dock is a transshipment facility at which trucks arrive with goods that must be sorted, consolidated with other products, and loaded onto outbound trucks bound for a retailer.

Cross docking is appropriate for products with large, predictable demands and requires that DCs be set up such that economies of scale in transportation are achieved on both the inbound and outbound sides.

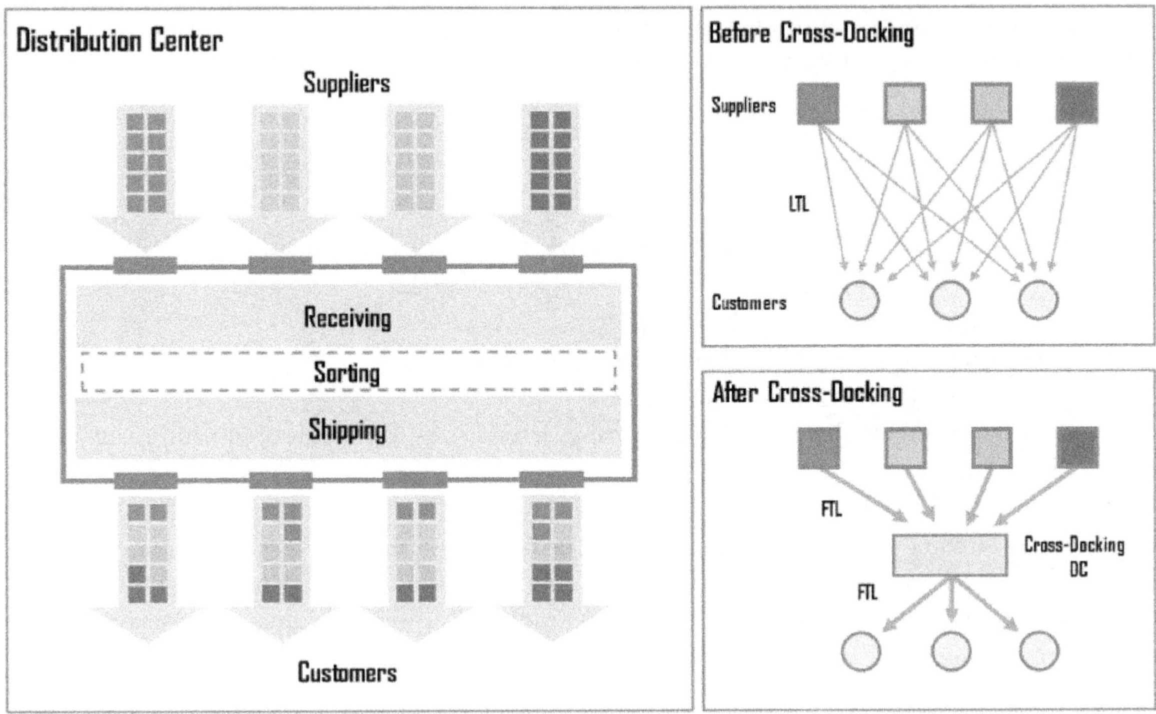

Figure 5: Shipments via central distribution center (Source: Dr. Jean Paul)

3.3.3 Shipping via DC using milk runs

Milk runs can be used from a DC if lot sizes to be delivered to each buyer location are small. Milk runs reduce outbound transportation costs by consolidating small shipments. The use of cross docking with milk runs requires a significant degree of coordination and suitable routing and scheduling of milk runs.

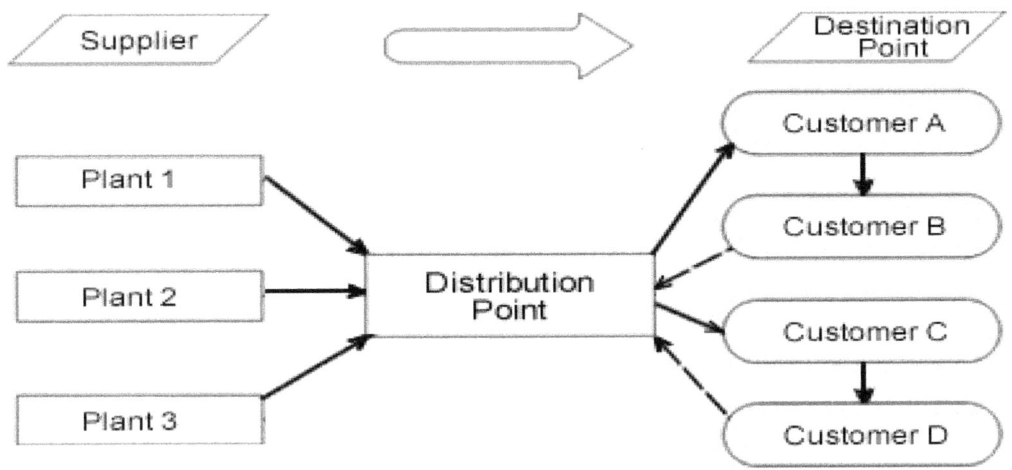

Figure 6: Shipping via DC using milk runs

3.3.4 Tailored Network

This option is a suitable combination of the previous options that reduces the cost and improves responsiveness of the value chain. Here transportation uses a combination of cross-docking, milk runs, and TL and LTL carriers, along with package carriers in some cases. The goal is o use the appropriate option in each situation. High demand carrier's products to high-demand retail outlets may be shipped directly, whereas low-demand products or shipments to low-demand retail outlets are consolidated to and from the DC. Operating a tailored network requires significant investment in information infrastructure to facilitate coordination.

3.4 Hub Advantages

- ➤ Hub consolidation reduces cost. Consolidation increases conveyance utilization
- ➤ Fewer conveyances are utilized
- ➤ Provides better level of service with fewer resources
- ➤ The relative distance is lower due to higher degree of circuitry
- ➤ Hub is more economical for smaller shipments
- ➤ Demand pattern is highly distributed due to many destinations from each origin and many origins from each destination
- ➤ It has good access for freight
- ➤ Hub acts as switching centers, consolidating intermediate cargo flows between multiple origins and destinations as well as contributing origin and destination traffic of their own.
- ➤ Customers may find the network to be simpler and more intuitive. Scheduling is more convenient for customers since there are fewer routes with more frequent service.
- ➤ Spokes are simpler, and new spokes can be connected rapidly.

3.5 Hub Disadvantages

- ➤ Cost of operating the hub is higher due to facility cost, handling cost that include unloading, loading and sorting, cost involved due to misrouting, damage or theft.
- ➤ Impact on service levels
- ➤ Productivity and utilization loss
- ➤ Congestion and delays
- ➤ Route scheduling is more complicated for the network operator. Scarce resources must be utilized carefully to avoid starving the hub. Careful traffic analysis and precise timing is required to keep the hub operating efficiently.
- ➤ The model is centralized and day-to-day operations may be relatively inflexible. Changes at the hub or even in a single route could have unexpected consequences across the network. It may be difficult or impossible to handle occasional periods of high demand between 2 spokes.
- ➤ The hub constitutes a bottleneck in the network. Total cargo capacity of the network is limited by the hub's capacity. Delays at the hub (e.g., weather) can result in delays across the entire network. Delays at a spoke (e.g., mechanical problems with an aircraft) can also affect the network, although to a lesser extent.

3.6 Performance Criteria

The performance criteria of the direct shipment with or without hub are influenced by number of trips, number of trucks, shipment cost and frequency of service.

Example I:

Pick up & delivery every day from pickup location to customers

Total demand of auto parts for 5 customers = 3TL delivery

Average distance between pickup location and customers = 1200 miles

Average distance from pickup location to hub = 600 miles

Average distance from hub to customers = 600 miles

Cost for transportation = $200 handling + $0.75 /mile

Cost for using hub = $150/day

<u>Direct Shipment</u>

Total no. of shipments = 5

Total handling cost = 5 X $200 = $1000

Total shipment cost = 5 X ($ .75 X 1200) = $4500

Total cost = $1000 + $4500 = $ 5500

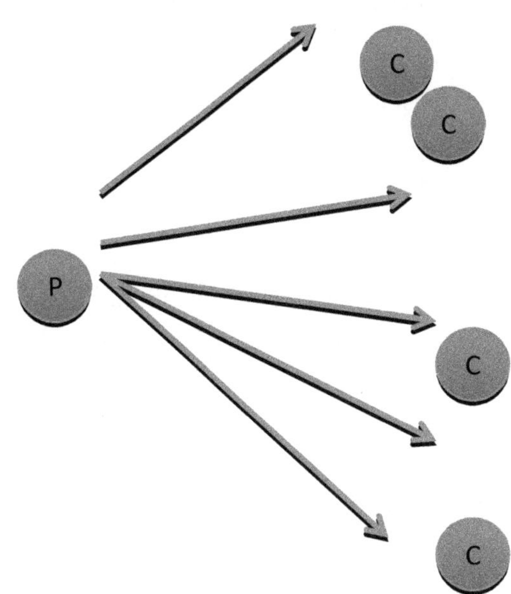

<u>Shipment using Hub</u>

Total no. of shipments $= 3 + 5 = 8$

Total handling cost $= 8 \times \$200 = \$1600$

Total shipment cost $= 8 \times (\$.75 \times 600) = \$3600$

Hub cost $= \$150$

Total cost $= \$3600 + \$1600 + \$150 = \$5350$

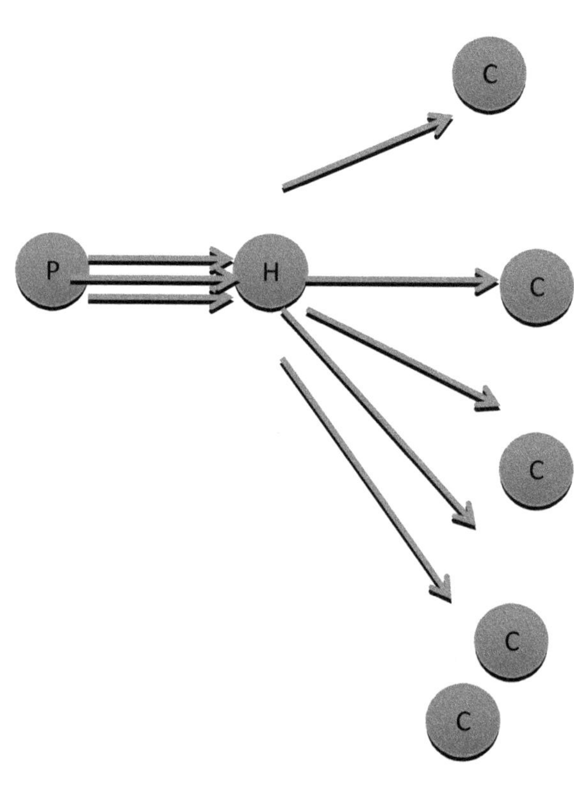

      The 3PL company should use the hub facility since the cost of transportation is lower with hub when compared to the direct network. The hub facility provides better service level as it needs to travel shorter distance and thus the customer demand can be fulfilled. The customer service level is improved by using hub facility.

Example 2:

Pick up & delivery every day from pickup location to customers

Total demand of auto parts for 5 customers $= 3$TL delivery

Average distance between pickup location and customers $= 1200$ miles

Average distance from pickup location to hub $= 600$ miles

Average distance from hub to customers $= 600$ miles

Cost for transportation $= \$250$ handling $+ \$0.75$ /mile

Cost for using hub = \$150/day

Direct Shipment

Total no. of shipments = 5

Total handling cost = 5 X \$250 = \$1250

Total shipment cost = 5 X (\$ .75 X 1200) = \$4500

Total cost = \$1250+ \$4500 = \$ 5750

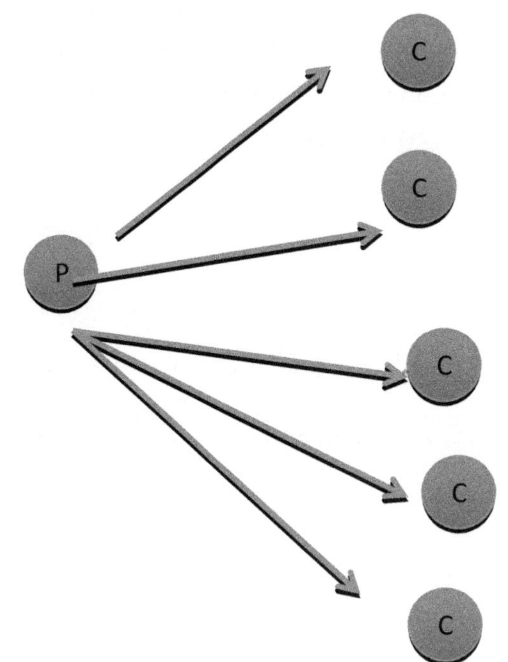

Shipment using Hub

Total no. of shipments = 3 + 5 = 8

Total handling cost = 8 X \$250 = \$ 2000

Total shipment cost = 8 X (\$ .75 X 600) = \$ 3600

Hub cost = \$150

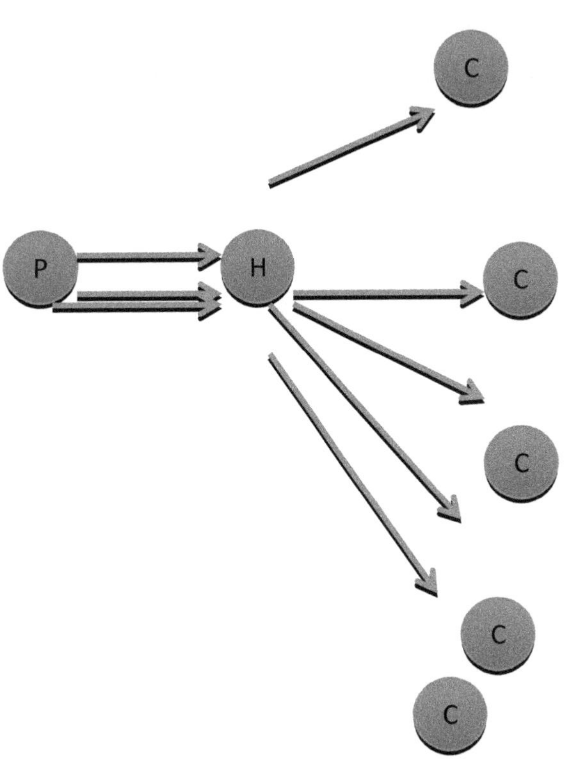

Total cost = $ 3600 + $ 2000 + $150 = $ 5750

      The 3PL company should use the hub facility since the cost of transportation is lower with hub when compared to the direct network. The hub facility provides better service level as it needs to travel shorter distance and thus the customer demand can be fulfilled. The customer service level is improved by using hub facility.

REFERENCES:

1. Robert Delaney, Industry impacts: Inventing and propelling the entire industries.
2. <http://logisticastillejo.wordpress.com/2010/04/13/transportation-networks/>

LECTURE 4

POWER POINT HANDOUTS

Transportation Infrastructure & Equipment

DR. MD SARDER

Transportation Infrastructure

➢The nodes and links associated with the transportation network are convenient for the purposes of network analysis and design, but these are not terms that are common when one talks with personnel involved in transportation management.

➢The network is generally considered to be infrastructure inasmuch as the links and nodes are, in most cases, in a fixed position and representative of significant long-term investment.

Transportation Equipment

➢Equipment is what operates on the network and provides the transportation service.

➢Like infrastructure, equipment represents assets, but it is movable and representative of less-significant and shorter-term capital investment.

➢Equipment is further classified as primary (e.g., truck) if it is used to directly provide the transportation service, or secondary (e.g., a fork lift) if it is used in support of the primary equipment.

Truck - Infrastructure

➢ The destination for the truck may be the shipper, the consignee, a warehouse or distribution center, a rail terminal, and an ocean terminal.

➢ A characteristic of large-sized DCs is that they are often located next to a high-capacity link.

➢The red band represents a large number of doors which allow trucks, which have backed up to the doors, to load and unload freight.

Distribution Center

Truck - Infrastructure

➢This dock happens to have three bays. Trucks back into position and the freight is unloaded into the consignee's receiving area.

Loading Dock

Truck - Infrastructure

➢Here is a 20-foot (TEU or Twenty-foot Equivalent Unit) in position at a dock.

➢Trucks that have a rail or ocean terminal as a destination generally pick-up or deliver containers.

Container on Chassis and Positioned at the Dock

Truck - Infrastructure

➢Here is a picture of a container chassis used for TEUs.

Container Chassis

Truck - Infrastructure

➢This terminal is an interface between truck and rail on the one hand, and ocean on the other.
➢Trucks enter and leave this ocean terminal through in and out gates. These trucks then pick-up or drop containers in the portion of the facility that is used for the storage of containers.
➢To the left one can see a container ship at berth where it is discharging and loading containers. The area between the ship & container storage looks to be a container staging area.

Long Beach Container Terminal

Truck - Infrastructure

> The principal links in truck transport range from complex super highways to country dirt roads.

Super Highway

Truck - Infrastructure

> Trucks of every size and shape are used to provide transportation service over every sort of highway and road imaginable (and some that can't be imagined).

Dirt Road

Truck - Equipment

➢Primary

▪To most of us, trucking equipment is defined by the 18 wheeler at one end of the spectrum.

18 Wheeler

Truck - Equipment

➢Secondary

▪Most trucks, such as shown in previous figure -18 Wheeler, are not loaded by manual means, but rather by means of the ubiquitous forklift.

▪The forklift comes along, places its tines into the slots that you see at the bottom of the pallet, the leifst the pallet slight above ground and moves it to where it need to go.

Palletized Freight on a Forklift

Water - Infrastructure

➢Water port provides services like loading, unloading, cross docking, temporary storage, connectivity, etc to shippers and carriers.

Port of Oakland, California, USA

Water - Infrastructure

➢The ocean is, of course, the link that ties the nodes together.

➢It is not just a matter of sailing off, but a rather more complicated selection of routes based on a number of factors. Some of these factors are predictable; channel depth, ship beam, surrounding ship traffic, demand from containerized service.

The Sea

Water - Equipment

➤Primary

'It's longer than the Eiffel Tower is tall, wider than the width of a football field and it can officially hold 11,000 20-foot-long shipping containers, though some suggest it can pack in even more.

Emma Maersk

Water - Equipment

➤Secondary

▪The side loader takes the container from a stack and lowers it onto a special chassis and power unit (hostler) combination that will then move the container elsewhere in the terminal.

Side loader and Hostler

Water - Equipment

> Secondary

- The crane is one of the most critical of the secondary pieces of equipment associated with the maritime mode. The crane will lift the container from the bomb cart and place it in position on the ship.

Ship and Crane

Water - Equipment

> Now the job the crane operator, if loaded the container aboard the ship, is to place the container in a cell position as called for by a load plan. You can see two empty cell positions to the right in the Figure.

Containers Aboard the Ship

Rail - Infrastructure

➤The principal node for the rail mode of transportation is the railway yard.

➤Railway yards are generally at the intersection of major railway lines (links) and serve the role of intramodal (rail to rail) interchange of railway cars. They also serve to facilitate the intermodal (e.g., rail to Ocean) interchange of shipments.

Railway Yard

Rail - Infrastructure

➤Track provides the links essential to rail mode of transportation.

➤Public or Private

Railway Track

Rail- Equipment

> The rail car is the asset that carries the freight for the customer.

> Here two 40-foot (FEU) containers are stacked in a well-car. The well-car lowers the center of gravity for the load and improves the clearance distances between the top of the containers and any structures above the rail.

Double-Stacked Containers in a Rail Well-Car

Rail - Equipment

> Railway cars are then coupled together into trains.

> The trains need to have power attached.

> Two locomotives are at the head of the train in this figure.

Railway Locomotives

Air - Infrastructure

➢Airport facilitates faster movement of goods. It works as a hub for companies such as UPS, FedEx.

➢Public or Private

Heathrow Airport Terminal 5

Air - Infrastructure

➢Sky is not unlimited!

➢Follow some specific routes.

➢The most flexible routes.

Sky Route

Air - Equipment

>Primary
- Passenger planes
- Cargo planes
- Chartered planes
- Public planes
- Private planes

Airplane on Runway

Air - Equipment

>Secondary
- Aviation cargo loader
- Conveyer belts
- Conveyer chains/plates

Aviation Cargo Loader

LECTURE 4

READING MATERIALS

4.1 Equipment

The infrastructure and equipment needed for each transportation system is a function of a number of variables. The challenge of Logisticians is to combine all the moving parts of a logistics system together in the most efficient manner. Therefore we must ask ourselves a number of questions prior to selecting the method of shipment.

Location of Origin and Destination
Where do goods need to be shipped?
How far will I be transporting the goods?

Type of Goods
What type of goods will be shipped?
How will the goods be packaged?

Once the details of the shipment have been determined we can analyze the equipment necessary for the successful completion of each shipment. There are two types of equipment necessary for completing each shipment of goods: Primary and Secondary.

Primary equipment is the defining factor of each mode of transportation. Primary equipment is considered planes, trains, trucks, and pipelines that provide the actual vessel for shipment. Each of these pieces of equipment come in a variety of forms and size is usually dependent on access to capital for investment.

Secondary equipment is the machines that are used in the loading/unloading and sorting processes within the transportation system. Secondary equipment varies in size and is highly dependent to the shipments which they handle. For instance, a forklift is a typical piece of secondary equipment used in truck transportation, while large maritime port operations require massive cranes to efficiently load and unload large container ships. Also, shipping containers, commonly known as TEU's, are considered secondary equipment because they are used to standardize shipments not to actually ship the goods.

4.2 Containerization

Transportation systems have become dependent on standardized shipping containers to ensure the timely and efficient transfer of containers between shipping modes. Goods that are shipped in standard metal containers of various sizes facilitate mechanized handling for each container. "In this way goods that might have taken days to be unloaded from a ship can now be handled in a matter of minutes (Slack 1998)." Mechanized handling allows the timely transfer between modes. There are many sizes of container available but the most prevalent container size is the 40 foot box, which can carry on average 22 tons of cargo with each haul. The beauty of these containers is that they are airtight, stackable, lockable, and can be outfitted with refrigeration for perishable loads. In the past, pallets were a common management unit in transportation, but their small size and lack of protective frame made transfer between modes of shipment labor intensive. The following chart lists the characteristics of the most popular container sizes. In the United States, a large

Figure I: Container Size

Container Size and Capacity	
International ISO Sizes (8.6' x 8')	
TEU (20 ft)	
Volume	1,140 cu.ft.
Payload	47,711 lbs
FEU (40 ft)	
Volume	2,390 cu. ft.
Payload	59,040 lbs
Domestic US (9' x 8.25')	
Length	53 ft
Volume	3857 cu. ft.
Payload	67,200 lbs

(element int'l 2011)

amount of domestic carriers utilize the 53 foot containers. Standardized shipping containers offer several advantages to shippers. Shipping containers are capable of shipping many types of products. A company could use the containers for raw material shipments, such as coal or aluminum, then immediately pack the same container with finished goods. Since the containers can ship anything companies do not need to invest as much capital in equipment. The shipment of container transportation also cost as much as twenty times less than regular bulk shipments. Additionally, containers that comply with ISO standards are capable of being handled around the globe by many different companies. As you can imagine, containers that can handled anywhere in the world greatly reduce the time and handling cost associated with any cargo. These containers can be stored anywhere and can be stacked, for storage purposes, multiple units high.

While containerized shipping does have many advantages it does have its downside. The unloading of shipping containers requires a minimum of 12 hectares. This land use requirement could potentially consume a large amount of terminal space, depending on the availability. The mechanized handling equipment mentioned earlier in the chapter, such as gantry cranes, yard equipment, etc. is expensive. These means the companies choosing to ship their products via standardized containers must initially commit to a large capital investment for all primary and secondary equipment. An additional consideration is that as much as 56% of their useful lives (10-15 years) idle or being repositioned. The return on investment for the containers may come quickly but the utilization ratio is generally considered quite low. While storage containers are capable of being stack for storage purposes, the stacking of these containers can be a complex problem. Containers must be stacked using the Last-In-First-Out (LIFO) only. Any other planning method would require a large amount time and money to pull a container from the bottom of a stack. It is true that the use of containerized shipping equipment has its advantages and disadvantages, but the job of the logistics manager is to determine if utilizing standard containers is worth the investment. And in many situations it is.

Lecture 5

Power Point Handouts

Transportation Rules & Regulations

DR. MD SARDER

Transportation Regulation

Transportation Rules & Regulations are for
- Vehicle operations
- Vehicle dimensions
- Safety of operators
- Safety of general public

Transportation Regulation

> Objective of this text
> > Provide insight into transportation's most important laws, rules, regulations, treaties and practices
> > Legal implications and consequences of logistical arrangements in the current deregulated environment

Creation of Transportation Laws

> Common Law – made by the state and federal courts, e.g., common career and its liabilities, page - 7
> Statutory Law – enacted by congress, e.g., trade embargos, page 7

Railroads

➢Railroad Rates – page 16

➢Railroads' Liability for loss, Damage and Delay – pages 20-21
  ➢Carmack Amendment – it requires railroads to assume liability for the full value of goods transported, railroads may limit their liability by providing a release rate..

Motor Carriers

➢Current Motor Carrier Regulation – page 32
  ➢What is a motor carrier?
  ➢Requirements of motor carrier

➢Cargo Insurance – pages 38 - 39
  ➢A requirement to operate cargo
  ➢A minimum of $5,000 per vehicle or $10,000 per incidence

Motor Carriers

➢Exemptions from Federal Motor Carrier Regulations:
  ➢Between Alaska and another state thru Canada
  ➢Rail/Truck/Water within terminal areas
  ➢Motor vehicle – school bus, taxi cabs, …
    ➢Pages 42-43

Motor Carriers

➢Motor Carrier Liability for Loss, Damage & Delay: page 85
  ➢Statutory liability – page 85
  ➢Suit filing time limits – min 2 years, page 111
  ➢Concealed loss & damage claims – page 113
  ➢Non-deliveries – page 114
  ➢Litigation issues – page 115
  ➢Arbitration & mediation – page 132

Shippers' & Carriers' Responsibilities

> Shippers' and carriers' duties and exposure to lawsuits: page 163
>> Shippers' responsibilities – pages 163 -169
>>> Packaging of the goods
>>> Duty to accept goods on delivery
>>> Liability for injury and death
>>> Accurate description of goods
>> Carriers' duties and responsibilities – pages 169 - 176
>>> Inspection of goods loaded by shipper
>>> Duties in loading and unloading
>>> Liabilities or damages for violating Interstate Commerce Act
>>> Driver regulations

Federal Motor Carrier Safety Regulations

> General qualifications of drivers - pages 555 – 556
> Disqualifications of drivers – pages 556 – 557
> Hours of service of drivers – pages 579 – 582
>> Maximum driving time
>> Record of duty status
>> Automatic on-board recording devices

Intermediary Regulation

- Brokers – pages 221 - 228
  - Regulation of brokers
  - Their liabilities
  - Brokers' insurance
- Freight forwarders – page 229
- Intermodal marketing companies – page 230
- Third party logistics provider – pages 230 - 236

Importing & Exporting Regulation

- Importing procedures – pages 237 - 240
  - Customs and importers
  - Entry of goods [also on pages 469 -470]
  - Customs examination
  - Importer obligations
- Foreign Trade Zones - pages 242 - 244

Importing & Exporting Regulation

- Exporting procedures – pages 244 - 246
  - Commerce Control List [also in pages 450-451]
  - Relevant Exporting Violations
  - Information required on shipper's export declarations [pages 436 -442]

HazMat Regulations

- HazMat – materials capable of posing unreasonable risk to health, safety and property according to Secretary of transportation, pages - 262 - 266
  - Identification
  - Classification
  - Packaging
  - Leveling
  - Emergency response
  - Training
  - Security requirements

Lecture 5

Reading Materials

5.0 Introduction

Congress has been enacting laws regulating various forms of transportation for over 100 years. Those laws are promulgated through the efforts of congregational committees having jurisdiction over transportation. The committees propose, investigate and draft laws, oversee their implementation by regulatory agencies, and propose new or remedial legislation to fine tune the law. The congressional committees having jurisdiction over transportation, as well as their names, may change from one Congressional session to next.

Transportation rules, regulations, and policies are made to enhance the efficiency of the transport system and enforce adherence to ideal code of conduct in the industry. There are several jurisdictions for federal and state governments, together with the agencies that operate herein.

Following the introduction, is the presentation of how transportation laws are created in chapter two. Chapter three presents the regulations in the railroad, airline and motor carrier industries. Chapter five entails an overview of some international transportation regulations, and a brief presentation of the HAZMAT regulations, while chapter six concludes the paper.

5.1 Creating Transportation Laws

There are two types of laws in the U.S, common law (case developed law) and statutory law.

Common Law: This is the body of law made by state and federal courts. An example of common law is the definition of a 'common carrier' as stated in 13 CJS Carriers S 3: 'A common carrier of goods is one who, as a regular business, undertakes to transport goods from place, offering its services to such as may choose to employ him and pay its charges.'

Occasionally, Congress codifies common law rules. For instance, Congress codified the common law rule of carrier liability when it enacted 49 U.S.C S 20 (11), (now S 11706 for rail road and S 1406 for motor carriers and freight forwarders), as Carmack Amendment to ICA, enacted as part of the Hepburn Act of 1906.

Statutory Law: Congress enacts statutory laws after hearings by the appropriate committees having jurisdiction over the subject matter. A key figure in enacting any law in the U.S is the chairperson of the committee. The chairperson controls hearing schedules, subjects for discussions, invitations to testify, and decides which bills be considered by the committee.

5.2 Federal Transportation Laws

Primarily federal statutes and treaties govern transportation of goods (property) in the interstate and foreign commerce or, if there is none directly on point, by federal common law. The federal authority to regulate transportation stems from the U.S Constitution, Article 1, Section 8, cl.3, which confers upon Congress the power to 'regulate commerce with foreign nations and among the several states.
The transportation of goods was the first form of interstate commerce subjected to federal regulation. Other forms of interstate commerce also subject to federal regulation, to a greater degree, are communications (telephone, telegraph, radio, television, etc.).

Congress enacts federal statues, which are published in the U.S Code (U.S.C). Congress delegates to federal agencies the power to formulate regulations necessary to implement and enforce the law, which are found in the Code of Federal Regulations (CFR). Proposed rulemakings are published in the Federal Register.

The first regulatory agencies created to implement the laws governing transportation were named the interstate Commerce Commission (ICC), having jurisdiction over surface transportation, the Civil Aeronautics Board (CAB), having jurisdiction over domestic airlines and air freight, and Federal Maritime Commission (FMC), having jurisdiction over water carriers and ocean freight forwarders.

Beginning in 1935, Congress required that truck lines obtain ICC approval before initiating or expanding interstate operations. While FMC was created to control ocean carriers rate cartels, to grant them antitrust immunity and to require strict adherence to their tariffs under the Shipping Act of 1916.CAB was also created in 1958 to regulate the operations of the airline industry. Regulations in the air industry were created to protect the carriers from excessive competition and ruinous financial results. These protectionist policies came under attack in 1977 which led to the airline industry being deregulated, followed by the deregulation of passenger travel in 1978. In 1985, the CAB was completely closed down, and on January 1, 1996, the ICC was abolished, with a significantly less powerful agency, the Surface Transport Board (STB), taking its place.

Regulations in the transportation sector have gone through a revolutionary trend from the start of its creation. For example, in the aftermath of September 11, 2001 terrorist attack on the U.S soil, the Congress created the Department of Homeland Security (DHS), charged with the responsibility of providing security on the homeland. This has led to the transfer of certain departments in the transport sector, like the Transportation Security Administration (TSA) and the National Coast Guards to the DHS. Border Patrol was also abolished with the creation of Border and Transportation Security (BTS).

When confronted with any transportation problem, one must understand the jurisdictional limits of the federal agencies having jurisdiction over transportation. The jurisdiction of the Secretary of Transportation, DOT, and the STB is defined by statue. The Secretary's duties and powers are generally described in 49 U.S.C. S 102. The general jurisdiction of the STB is described in 49 U.S.C. S 10501 (for rail), S 13501 (for motor carriers), S 13521 (for water carriers) and "13531 (for domestic freight forwarders). A party desiring to invoke the specific jurisdiction of the STB over rates and practices must determine whether a 'user fee' will be charged. 49 C.F.R S 1002.3.

5.3 Railroad Industry Regulations

The enactment of the Act to Regulate Commerce by the congress in 1887 was as a result of financial manipulations of the owners of railroads, discriminatory pricing by their operators and disaffection by the users which are principally farmers trying to transport their goods to the markets. The act was short and to the point. In effect, it imposed only a handful of requirements on the railroads. Their prices have to be just and reasonable, non-discriminatory and non-preferential. Railroads rates had to be published and any deviation from these tariff rates was illegal. This requirements was known as 'the filed rate doctrine'.

The railroads were nationalized during World War I. Although ownership remained in private hands, their operations were delegated to the Director General, a Government appointee. Whether because of the Director General's mismanagement or the heavy demands placed upon railroads as a result of the war's transportation requirements, the railroads emerged deplorable from the Director General's control. Congress responded by enacting the Transportation Act of 1920 and its provision marked the pinnacle of the pervasive regulation to which the railroads were to remain subject for approximately the next sixty years. The Act of 1920 carried forward the provisions of the Act to Regulate Commerce of 1887, augmented in the meantime by the authority of the ICC to prescribe the rates to be assessed by the railroads on specified commodities or in designated traffic lanes and to award reparations to shippers for railroad rates found to have been assessed unlawfully. The 1920 Act provided that railroads could neither construct, discontinue service on, or abandon railroad lines without the advance approval of the ICC, upon its determination that the public convenience and necessity required the proposed action. The ICC was charged with the task of devising a plan for the consolidation of the railroads into a limited number of regional systems, and railroads could merge in furtherance of that ICC plan.

As amended in the Transportation Act of 1920, the ICA treated the railroads as if they were public utilities. The difficulty, however, was that no sooner had the 1920 legislation been enacted than the railroads began to witness excessive competition from truck and bus transportation. Therefore, the sought to stem the erosion of their traffic by supporting the enactment of the Motor Carrier Act of 1935, which imposed upon motor carriers of freight and passengers a regulatory scheme akin to that of the railroads.

Nominally, at least, the STB continues to have jurisdiction over the nation's rails roads. Its jurisdiction, however is quite limited. The STB does not have authority to regulate propriety railroads – those that serve a single industry and do not hold themselves out to serve the public.

There are a number of reasons why a shipper may desire railroad service. Sometimes, as, for example, in the transportation of bulk chemicals or plastics, it costs less compared to trucking. Extraordinary heavy items are way easier to ship by rail too especially if the shipper is located just as close to an existing rail.

Essentially, there is no regulation of railroads rates. As already noted, much of the railroad transportation has been declared exempt. Railroad rates on those exempt commodities and services may be set by the railroads at their discretion, without even the pretence of rate supervision by the STB. In theory, an aggrieved shipper may petition the STB to vacate exemption to entertain a complaint that the rate assessed on the exempt commodity exceeds a reasonable level. Although, such relief is more illusory than real. Some measure of protection was sought to be afforded shippers of agricultural products, including grain. Summaries of such contracts must be filed with STB, and a railroad can enter into contracts for the transportation of agricultural products utilizing not more than 40 percent of its cars.

One of the most important and most frequently litigated statutes is the one governing railroads liability for loss, damage and delay to goods in their possession, called the 'Cormack Amendment'. The version currently applicable to rail carriers appears at 49 U.S.C. S 11706. In view of the importance of this subject, a more detailed discussion of its background, terms and application of the liability standard to rail traffic warranted.

5.4 Motor Carrier Regulation

There was a turnaround in the filing of motor carrier tariffs with the ICC in 1994. Effective January 1, 1996, the ICC was 'unsettled', or legislated out of existence by ICCTA, and most of the former regulatory controls over motor carriers ended. Remaining features of regulations were transferred to the Secretary of Transportation, to the Department of Transportation (DOT), the Federal Highway Administration (FHWA), whose powers were transferred later to the Federal Motor Carrier Safety Administration (FMCSA), or to the newly created Surface Transportation Board (STB). The jurisdiction of the Secretary of Transportation over the interstate transportation, which encompasses the DOT, the FMCSA and the STB, is provided in 49 U.S.C S13501.

The term 'motor carrier' is defined in the ICCTA as a 'person providing motor vehicle transportation for compensation'. 49 U.S.C. S13102. The more recent Act does not separately define motor common carrier or motor contract carrier, the repercussions of which are debatable. In addition, a motor carrier shall 'provide safe and adequate service, equipment, and facilities'. 49 U.S.C S14101. There are requirements for those who wish to gain motor carrier status and have authority to operate as such from the federal government. They include:

    i.   File an application with the Federal Motor Carrier Safety Administration (FMCSA)

    ii.   File evidence of the proper insurance

    iii.   Designate an agent for the service of legal process and

    iv.   Comply with the FMCSA's safety regulations

State rules governing corporations typically provide that a corporation desiring to do business within the state must designate an agent in the state to receive service of process on their behalf. Those agents are known as process agents.

It is also pertinent to mention that certain commodities, geographic zones, and equipment fall outside the scope of the federal regulation, commonly referred to as 'exemptions'. The exemptions include:

a. Transportation by motor vehicle between Alaska and another state through Canada, 49 U.S.C S 13502

b. Transportation by rail, truck or water carriers within terminal areas incidental to interstate or foreign commerce, 49 U.S.C S 13503

c. Intrastate motor carrier operations within Hawaii (except for household goods), 49 U.S.C S 13504

d. Motor vehicle transportation by a person in furtherance of its primary business, including subsidiaries in which the parent owns a 100 percent interest, commonly known as 'private carriage' 49 U.S.C S 13505

e. Motor vehicle transportation, as described in 49 U.S.C S 13506 (a);

5   School vehicles

6   Taxicabs

7   Hotel vehicles

8   Farmers agricultural commodities

9   Agricultural cooperatives vehicles

10 Motor vehicles hauling livestock, unmanufactured agricultural or horticultural commodities, fishery products, feed, seeds or plants

11 Newspapers

12 Ground movements prior or subsequent to a movement by an air carrier, or in lieu of air because of weather conditions, mechanical failure, or due to conditions beyond the control of the carrier or the shipper

13 National park vehicles

14 Commutation vehicles carrying not more than 15 passengers daily

15 Used pallets and empty shipping containers and devices

16 Natural, crushed, vesicular rock used for decorative purposes

17 Wood chips

18 Passenger brokers and

19 Broken, crushed or powdered glass.

Fines and penalties for motor carriers, brokers or freight forwarders violation of the interstate Commerce Act are provided in 49 U.S.C S 14901-14914.

5.5 Regulations in the Airline Industry

Federal regulations of the airline industry began in 1938 with congress enactment of the Civil Aeronautics Act, which led to the establishment of the Civil Aeronautics Board (hereinafter 'CAB'). The Federal Aviation Act of 1958 vested all regulatory authority over safety in the newly created Federal Aviation Administration (FAA). The airlines were held to 'no negligence' of accountability until 1972, when an international treaty on air transportation of goods, the Warsaw Convention, directed that a due diligence standard be adopted. In 1977, strict liability was ordered as the standard of liability for domestic air carriers.

During its initial stages, the CAB did not intervene in carrier loss and damage liability matters. However, in 1969 it was forced to launch an investigation into liability, in response to complaints from shippers of flowers, seafood and other goods that the airline industry claim practices and tariff limitations of liability for loss, damage or delay were unreasonable. Also, the deregulation of airline industry began in 1977 with deregulation of air cargo traffic. As a result of deregulation, air cargo rates and charges were no longer filed with or regulated by the CAB. Although airline rates, rules, liability terms and services are currently deregulated, airline tariffs continue to be published in a tariff publication entitled "Airline Tariff Publishing Company" (ATCP). However, effective January 1, 1983, airlines are no longer required to file their tariffs with any federal agency.
It is federal common law that governs liability issues in interstate air shipments. Under federal common law, an airline must comply with the released value doctrine if it wishes to limit its liability. That doctrine allows a carrier to limit its liability in exchange for offering a lower rate to shipper. The carrier must provide the shipper with;

i.   Reasonable notice of limited liability

ii.   A fair opportunity to purchase a higher level of liability.

5.6 Importing Procedures

a. Customs and Importers

The Customs and Border Patrol (CBP), sets out the following as its mission;

1. Protecting the nations revenue by assessing and collecting duties, taxes, and fees incident to international traffic and trade

2. Controlling, regulating, and facilitating the movement of carriers, people, and commodities between the U.S. and other nations

3. Protecting domestic industry and labor against unfair foreign competition

4. Protecting American consumer and environment form hazardous products

5. Detecting, interdicting and investigating fraudulent smuggling and other illegal practices aimed at prohibited articles entering into the U.S.

6. Detecting interdicting and investigating fraudulent activities intended to avoid payment of duties, taxes, or fees, or to otherwise evade legal requirements to international trade

7. Detecting, interdicting, and investigating illegal international trafficking in arms, munitions, currency, and acts of terrorism at the U.S. ports of entry.

The Customs Modernization Act governs the relationship between importers and customs in the U.S. through a principle known as "Informed Compliance". The phrase refers to the shared responsibility between importers and customs to foster more efficient and expedited clearance of goods through U.S. borders.

b. Entry of goods

For entry of goods into the U.S., an importer must file entry documents for the goods with a port director at the goods port of entry. The importer of record has a duty to prepare the goods for inspection and release, and must use reasonable care doing so. Legal entrance in the U.S. requires that goods, after arrival at a U.S. port of entry, be authorized for delivery by Customs and that estimated duties and taxes are paid. A proper documentation of imports must also be done to determine whether the goods may be released from Customs custody.

c. Customs Examination of Entry Goods/Documents

Customs examine goods upon entry to the U.S. to;

1. Determine the value of goods for their dutiable status and other custom purposes

2. Determine whether goods must be marked with their country of origin or require special labelling

3. Detect prohibited and prevent entrance

4. Determine if goods are correctly invoiced

5. Determine if goods exceed invoiced quantities or if a shortage exists.

d. Importer Obligations

An importer is expected to exercise reasonable care in carrying out all activities related to importation of goods. Some of these obligations include;

1. Providing a complete and accurate description of goods

2. Providing a correct tariff classification

3. Obtaining a Customs ruling on the description, marking, country of origin, and valuation of goods and its tariff classification

4. When claiming goods are entitled to a conditionally free or special tariff classification, assuring they qualify for such status

5. Providing a proper declared value

6. Providing the correct country of origin marking (if required)

7. If applicable, establishing the legal right to import goods that are trademarked or copyrighted.

8. Assuring goods comply with other relevant agency requirements (FDA, EPA, FTC, DOT, etc.)

9. Compliance with Commerce Department dumping or countervailing duty investigations

10. Filing the correct type of Customs entry (consumption, mail, etc.)

11. Compliance with any special regulations that may apply to the commodity (textiles, hazardous materials, perishable goods etc.)

It is advisable that importers seek expert assistance such as lawyers, accountants, custom brokers and custom consultants to ensure proper compliance.

e. Penalties

Any person, who fraudulently or negligently enters, introduces, or attempts to enter or introduce goods into the U.S. in a way that violates U.S. Customs laws and regulations may be subject to criminal and civil penalties. Importers goods may be seized or forfeited to pay the penalty. Criminally, an importer can be subject to up to two years imprisonment, a fine, or both.

5.7 North American Free Trade Agreement (NAFTA)

Implementation of the North American Free Trade Agreement (NAFTA) began on January 1, 1994. This agreement will remove most barriers to trade and investment among the United States, Canada, and Mexico. Under the NAFTA, all non-tariff barriers to agricultural trade between the United States and Mexico were eliminated. In addition, many tariffs were eliminated immediately, with others being phased out over periods of 5 to 15 years. This allowed for an orderly adjustment to free trade with Mexico, with full implementation beginning January 1, 2008.

The agricultural provisions of the U.S.-Canada Free Trade Agreement, in effect since 1989, were incorporated into the NAFTA. Under these provisions, all tariffs affecting agricultural trade between the United States and Canada, with a few exceptions for items covered by tariff-rate quotas, were removed by January 1, 1998.Mexico and Canada reached a separate bilateral NAFTA agreement on market access for agricultural products. The Mexican-Canadian agreement eliminated most tariffs either immediately or over 5, 10, or 15 years. Tariffs between the two countries affecting trade in dairy, poultry, eggs, and sugar are maintained. Chapter 52 provides a procedure for the interstate resolution of disputes over the application and interpretation of NAFTA. It was modelled after Chapter 69 of the Canada-United States Free Trade Agreement.

5.8 Hazardous Materials Regulations (HMR)

The Hazardous Materials Regulations (HMR; 49 CFR Parts 171-180) specify requirements for the safe transportation of hazardous materials in commerce by rail car, aircraft, vessel, and motor vehicle. These comprehensive regulations govern transportation-related activities by offertory (e.g., shippers, brokers, forwarding agents, freight forwarders, and warehouses); carriers (i.e., common, contract, and private); packaging manufacturers, reconditioners, testers, and re-testers; and independent inspection agencies. The HMR apply to each person who performs, or causes to be performed, functions related to the transportation of hazardous materials such as determination of, and compliance with, basic conditions for offering; filling packages; marking and labelling packages; preparing shipping papers; handling, loading, securing and segregating packages within a transport vehicle, freight container or cargo hold; and transporting hazardous materials.

In general, the HMR prescribe requirements for classification, packaging, hazard communication, incident reporting, handling and transportation of hazardous materials. The HMR are enforced by RSPA and DOT's modal administrations: the FAA, the Federal Highway Administration (FHWA), the Federal Railroad Administration (FRA), and the United States Coast Guard (USCG). Federal law provides for civil penalties of not more than $25,000 and not less than $250 for each violation. An individual who wilfully violates a provision of the HMR may be fined, under Title 18 U.S.C., up to $250,000, be imprisoned for not more than 5 years, or both; a business entity may be fined up to $500,000.

The information presented in this document highlights some of the requirements of the HMR which can affect transportation safety, but does not address many of the specific provisions and exceptions contained in the HMR. This advisory notice is intended to provide general guidance. It should not be used as a substitute for the HMR to determine compliance.

5.8 Foreign Trade Zones (FTZ)

A Foreign-Trade Zone is a secure area defined by US law as being outside the Customs Territory of the United States; no duty is paid and certain state, local and federal taxes are eliminated on foreign goods or material brought into the Zone, until the goods are "entered" from the zone into U.S. customs territory. A Foreign-Trade Zone is associated with an air or sea port of entry. In the United States, the Foreign-Trade Zones Board of the U.S. Department of Commerce regulates Foreign Trade Zones. In the rest of the world, such zones are often called "Free Trade Zones."

Foreign goods can be shipped directly from an overseas port into a Foreign Trade Zone where they will be "admitted" into the Zone rather than "entered" into U.S. Customs Territory. Once admitted into a Foreign-Trade Zone, goods may be held for storage or they may be labelled, repackaged, processed, assembled, or manufactured into a finished product.

The Zone is physically on U.S. soil, employs U.S. labor, and within certain limits, goods produced in the Zone can be labelled "Made in the USA". Suffice it to say that if a product can properly be labelled "Made in the USA" outside of an FTZ, then it can also be labelled as such if manufactured inside an FTZ. The completed or repackaged products can then be "entered" from the Zone into the United States, or they can be shipped back overseas without ever having entered U.S. Customs Territory. If a finished product manufactured in the Zone is entered into Customs Territory, the importer pays duty only on the foreign content of the product manufactured in the Zone. The rate of duty on the finished or repackaged product may be lower than the duty that would have been paid on the imported components or raw materials had they been imported without Zone use.

The Zone's value to the user is the reduction in or the deferral of import duty on goods with high tariff rates. Also inventory in a Zone is exempt from federal excise taxes and state or local ad valorem taxation. The value of a Foreign-Trade Zone to the U.S. economy is the Zone's ability to create jobs through increased international trade and to prevent irrational tariff rates from causing U.S. manufacturers to relocate overseas.

Conclusion

Transportation rules, regulations, and policies are made to enhance the efficiency of the transport system and enforce adherence to ideal code of conduct in the industry. There are several jurisdictions for federal and state governments, together with the agencies that operate herein. Congress enacts federal statues, which are published in the U.S Code (U.S.C). Congress delegates to federal agencies the power to formulate regulations necessary to implement and enforce the law, which are found in the Code of Federal Regulations (CFR). Proposed rulemakings are published in the Federal Register.

Side-by-side, transportation rules, regulations and policies in the U.S has been evolving, with improvement in technology and the transport systems. It has also seen a radical shift because of global security situation since 911. Policies have been made not only for efficiency reasons but also to enhance the security of the homeland.

References

Anthony, A.r., 1992. Interpretive Rules, Policy Statements, Guidances, manuals, and the Like- should Federal Agencies use Them to Bind Public. Duke Law Journal.

Jefferson, T., 2013. *Hazardous Materials Transportation Policy and Procedures.* U.S. Department of Energy.

Nolaba http://nolaba.org/wp-content/uploads/2013/02/Foreign-Trade-Zone-White-Paper.pdf Retrieved on 05/03/2013

USDA (2013). Foreign Agricultural Service. http://www.fas.usda.gov/itp/policy/nafta/nafta.asp Retrieved 03/04/2013

Lecture 6
Power Point Handouts

Intermodal Transportation

DR. MD SARDER

Intermodal Transportation

Intermodal Transportation

➤It should be evident by now that each of the modes, and its accompanying infrastructure and equipment, have strengths and weakness relative to the nature of the service desired. As one moves from local services through regional, national, and international services, one is tempted to deploy those modes on that section of the service where their strengths prevail and their weaknesses can be overcome.

➤The idea being that by combining the modes – intermodal – one can provide a level of service that satisfies the customer at a cost less than using a single mode. Of course, in some cases, a particular modal opportunity does not exist.

Intermodal Transportation

➢Intermodal service is about combing the modes, the networks, the infrastructure and equipment into an optimally priced integrated service offering, and managing that offering in a seamless fashion, such that provides a competitive advantage to the customer as well as the service provider.

➢"The concept of transporting passengers and freight on two or more different modes in such a way that all parts of the transportation process, including the exchange of information, are efficiently connected and coordinated."

Gerhardt Muller, Eno Transportation Foundation

Intermodal Transportation

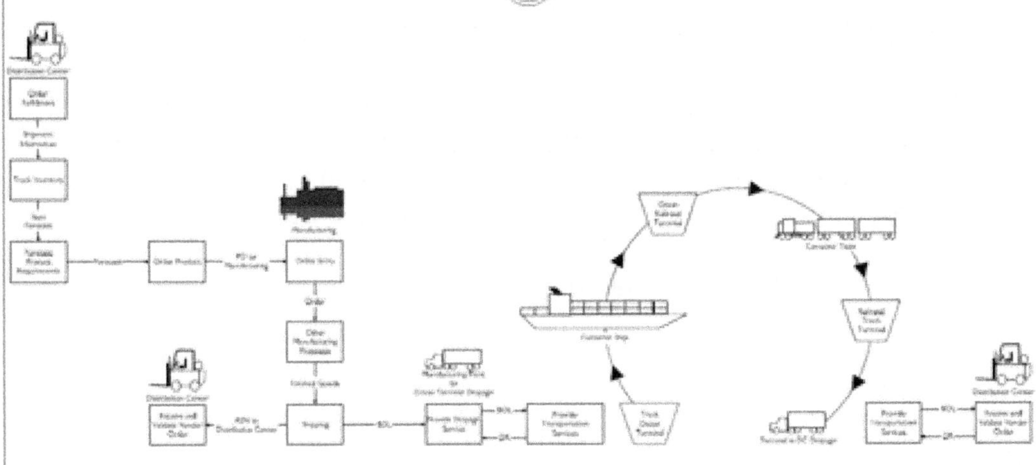

A Hypothetical Example

Intermodal Transportation

➢The efficiency of intermodal rests on two key principles.

➢The Physical Interface: The ease and speed with which the freight is transferred from mode to mode with no loss of or damage to the freight in the transfer process.

➢The Information Interface: The ease and speed with which the freight is transferred from mode to mode is dependent upon the ease and speed with which the information is transferred from mode to mode. As a rule, freight should not move without its associated information.

Intramodal Transportation

➤ *"Transfer made between vehicles of the same mode."*

➤The movements of passengers or freight within the same mode of transport. Although "pure" intramodal transportation rarely exists and an intermodal operation is often required (e.g. ship to dockside to ship), the purpose is to insure continuity within the network.

➤Also known as "Transmodal Transportation"

Multimodal Transportation

➤"The carriage of goods by at least two different modes of transport."

➤A set of transport modes offering connections between a set of origins and destinations. Although intermodal transportation is possible, it does not necessarily occur.

Intermodal Transportation

> Limitations:
> - Capital intensive
> - Demands close coordination and cooperation among participants
>
> Key Issues:
> - Standardization of equipment
> - Sophisticated communications

Intermodal Equipment

> In the past, pallets were a common management unit, but their relatively small size and lack of protective frame made their intermodal handling labor intensive and prone to damage or theft.

> By the early 1930s about three days were required to unload a rail boxcar containing 13,000 cases of unpalletized canned goods. With pallets and forklifts, a similar task could be done in about four hours.

Palletized Freight on a Forklift

Intermodal Equipment

➢Better techniques and management units for transferring freight from one mode to another have facilitated intermodal transfers.

➢Early examples include piggyback (TOFC: Trailers On Flat Cars), where truck trailers are placed on rail cars, and LASH (lighter aboard ship), where river barges are placed directly on board sea-going ships. Double stacked containers are known as COFC: Containers On Flat Cars.

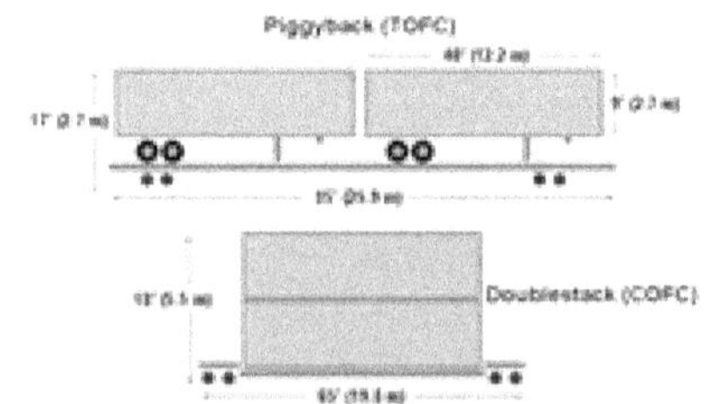

Double-Stacked Containers in a Rail Well-Car

Intermodal Equipment

➢The major development undoubtedly has been the container, which permits easy handling between modal systems. Containerized traffic has surged in recent years, underlining its adoption as a privileged mean to ship products on international and national markets.

➢Containers have become the most important component for rail and maritime intermodal transportation.

➢Containers permit the mechanized handling of cargoes of diverse types and dimensions that are placed into boxes of standard sizes. In this way goods that might have taken days to be loaded or unloaded from a ship can now be handled in a matter of minutes (Slack 1998).

Containerization

> A large standard size metal box into which cargo is packed for shipment aboard specially configured transport modes. It is designed to be moved with common handling equipment enabling high-speed intermodal transfers in economically large units between ships, railcars, truck chassis, and barges using a minimum of labor.

> The container serves as the unit load and provides complementarily between freight transportation modes.

Containerization

- ❖ Characteristics
  - Airtight, Stackable, Lockable
- ❖ International ISO Sizes (8.6' x 8')
  - TEU (20 ft)
    - Volume 33 M3
    - Total Payload 24.8 kkg
  - FEU (40 ft)
    - Volume 67 M3
    - Total Payload 28.8 kkg
- ❖ Domestic US (~9' x 8.25')
  - 53 ft long
    - Volume 111 M3
    - Total Payload 20.5 kkg

Containerization

➢There are many sizes of containers available but the most prevalent container size is however the 40 foot box, which in its 2,400 cubic feet which carry on average 22 tons of cargo.

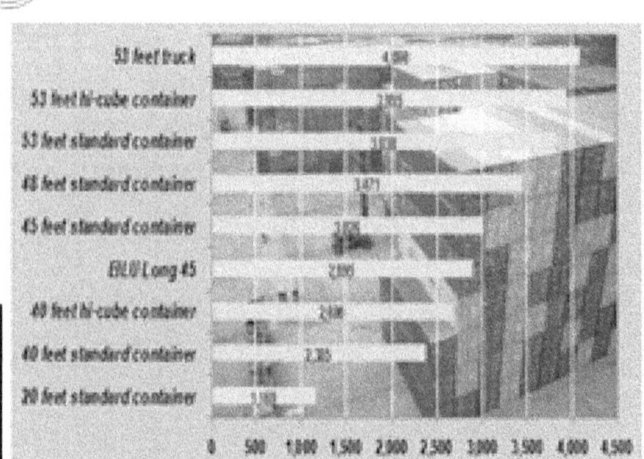

Carrying Capacity of Containers
Source: adapted from Robert C. Leachman (2005)

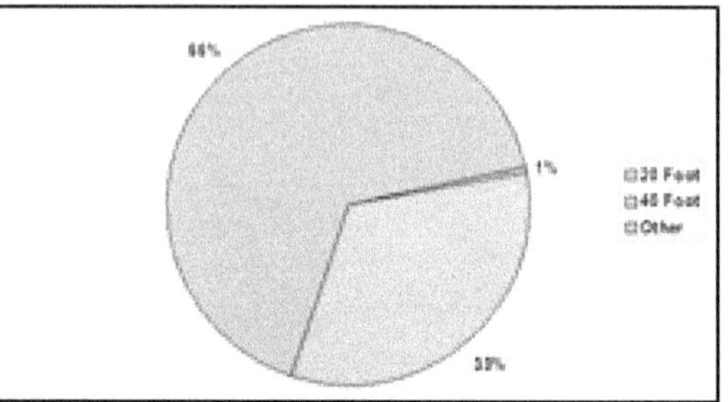

Global Fleet of Containers, 2000
Source: UNCTAD (2000) Review of Maritime Transport

Containerization

➢In the United States, a large amount of domestic containers of 53 foot are also used. Containers are either made of steel (the most common for maritime containers) or aluminum (particularly for domestic) and their structure confers flexibility and hardiness.

53" Fleet Domestic Containers
Source: Photo: Dr. Jean-Paul Rodrigue

US Containerized Trade with Asia, 1996-2007
Source: UNCTAD, Review of Maritime Transport.

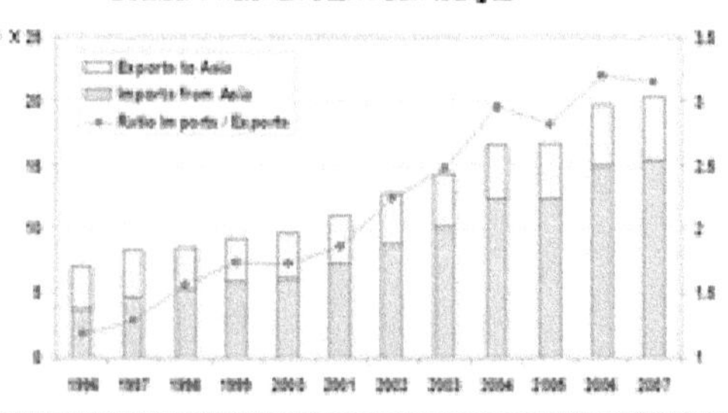

Advantages of Containers Use

> **Flexibility of usage:** It can transport a wide variety of goods ranging from raw materials (coal, wheat), manufactured goods, and cars to frozen products. There are specialized containers for transporting liquids and perishable food items in refrigerated containers called "reefers".

> **Costs:** Relatively to bulk, container transportation reduces transport costs considerably, about 20 times less.

20-Foot Tank Containers
Source: Photo courtesy of Gary and Matt Hannes, The Intermodal Container Web Page

Advantages of Containers Use

> **Standard transport product** : A container can be manipulated anywhere in the world as its dimensions are an ISO standard.

> **Speed:** Transshipment operations are minimal and rapid. A modern container ship has a monthly capacity of 3 to 6 times more than a conventional cargo ship. This is notably attributable to gains in transshipment time as a crane can handle roughly 30 movements (loading or unloading) per hour. Port turnaround times have thus been reduced from 3 weeks to less than 24 hours.

> **Warehousing:** The container limits the risks for goods it transports because it is resistant to shocks and weather conditions. The packaging of goods it contains is therefore simpler and less expensive.

> **Security:** The contents of the container are anonymous to outsiders as it can only be opened at the origin, at customs and at the destination.

Disadvantages of Containers Use

➤ **Site space requirements**: Containerization implies a large consumption of terminal space. A containership of 5,000 TEU requires a minimum of 12 hectares of unloading space.

➤**Infrastructure costs:** Container handling infrastructures, such as gantry cranes, yard equipment, road and rail access, represent important investments for port authorities and load centers.

➤ **Stacking:** The arrangement of containers, both on the ground and on modes is a complex problem. LIFO only.

➤**Empty travel:** On average containers will spend about 56% of their 10-15 years lifespan idle or being repositioned empty. Either full or empty a container take the same space on ship, truck or train.

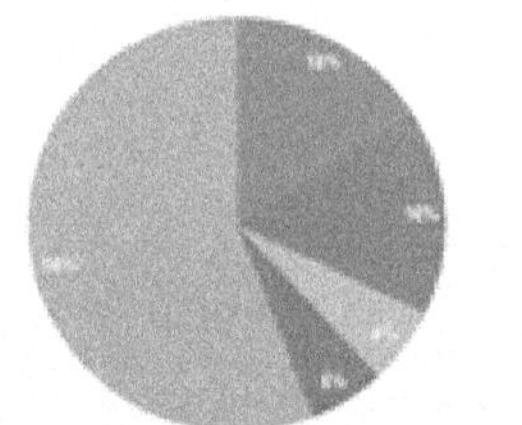

Container Usage during its Life-Span
http://www.interasset.com/docs/AssetManagementWP.pdf

Intermodal Transportation Costs

➤There is a relationship between transport costs, distance and modal choice that has for long been observed. It enables to understand why road transport is usually used for short distances (from 500 to 750 km), railway transport for average distances and maritime transport for long distances (about 750 km).

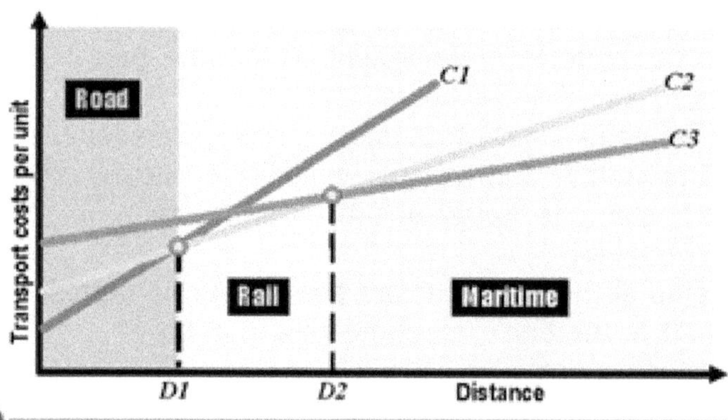

Distance, Modal Choice and Transport Cost
D1 = 500 -750 KM
D2 = Near 1500 KM
Source: Photo: Dr. Jean-Paul Rodrigue

Intermodal Transportation Costs

➢The concept of economies of scale applies particularly well to container shipping.
➢It is thus not surprising that maritime shipping companies have introduced larger and larger containerships, particularly over long distance routes.

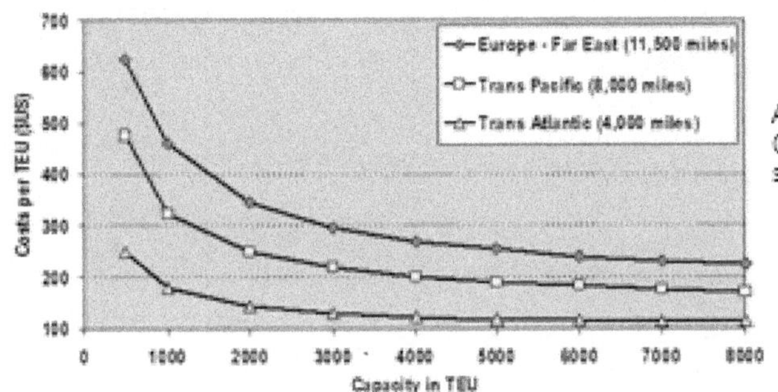

Average Cost per TEU by Containership Capacity and By Route, 1997
Source: Cullinane, K. and M. Khanna

Intermodal Transportation Costs

➢While maritime container shipping companies have been pressing for larger ships, transshipment and inland distribution systems have tried to cope with increased quantities of containers.
➢The growth in capacity comes with increasing problems to cope with large amounts of containers to be transshipped over short periods of time as shipping companies want to reduce their port time as much as possible.

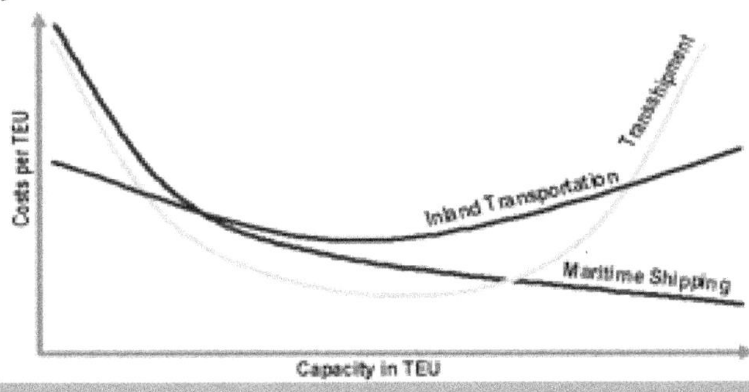

Economies and Diseconomies of Scale in Container Shipping
Source: Photo: Dr. Jean-Paul Rodrigue

Case Study

> **Shipping Shoes from China**

How should I ship my shoes from Shenzhen, China to Kansas City, USA?

- Shoes are manufactured, labeled, and packed at plant
- ~4.5M shoes shipped per year from this plant
- 6,000 to 6,500 shoes shipped per container (~700-750 FEUs / year)
- Value of pair of shoes ~$35

Case Study

40 shipping lines visit these ports each w/ many options
Examples:
- APL - APX-Atlantic Pacific Express Service
  - Origins: Hong Kong -> Kaohsiung, Pusan, Kobe, Tokyo
  - Stops: Miami (25 days), Savannah (27), Charleston (28), New York (30)
- CSCL - American Asia South loop
  - Origins: Yantian -> Hong Kong, Pusan   Stops: Port of Los Angeles(16.5 days)

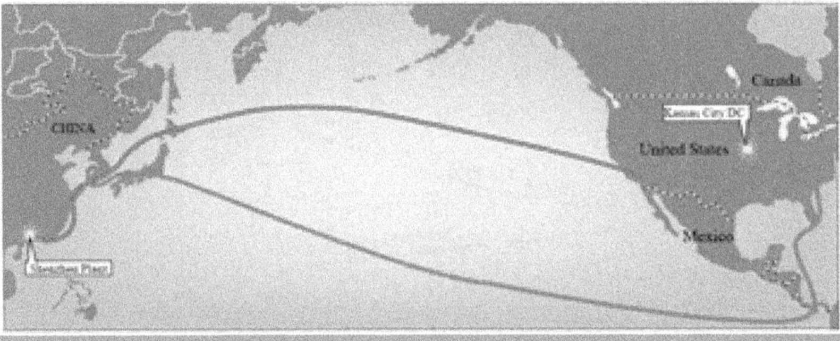

Case Study

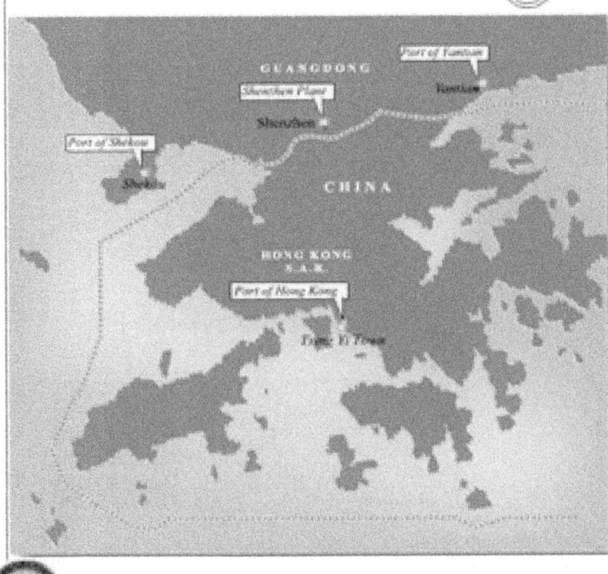

Inland Transport @ Origin

3 Port Options

- Shekou (30k)
  - Truck
- Yantian (20k)
  - Rail
  - Truck
- Hong Kong (32k)
  - Rail
  - Truck
  - Barge

In Hong Kong
- 9 container terminals

Case Study

Inland Transportation in US

Case Study

Port of New York / New Jersey
❖ Maher Terminal

- Express Rail II NS RR
  - Double stack thru:
  - Harrisburg, Pittsburgh, Cleveland, Ft. Wayne, to Kansas City
- CSX RR (5-10 days)
  - Double stack thru:
  - Philadelphia, Baltimore, Washington, Pittsburgh, Stark, Indianapolis, to Kansas City
- Truckload (2.5 - 3 days)
  - NJ Turnpike to I-78W, I-81S, I-76/70 to Kansas City

Case Study

Transport Options

- So how do I ship shoes from Shenzhen to Kansas City?

- What factors influence my decision?

- Consider different types of networks

  - Physical
  - Operational
  - Strategic

Lecture 6

Reading Materials

6.1 Modality

These modes of transportation may be combined in any way imaginable but the task of the logistician is to make the decisions that will ultimately combine these five modes of transportation in the most efficient and economical fashion. This section will highlight a number of common combinations used in today's global supply chain.

***6.1.1 Unimodal* –** Unimodal refers to the shipment of goods from origin to destination using only one mode of transportation. This particular method is most often used in road transport because the vast majority of shipments by other modes of transportation would require multiple forms of shipment.

6.1.2 *Bimodal* - This method of freight shipment combines two modes of transportation. Combining two forms of transportation is more common than the Unimodal method previously discussed. This method is most

useful for long distance shipments which cross multiple state or country borders. Recent developments in technology have actually combined two forms on transportation into a single freight carrier, similar to TEU containers. This technology is an example of assets that create efficiency through form and function.

6.1.3 *Intermodal* – The intermodal method of freight shipment has become the most effective way of bringing goods to varied markets throughout the world simultaneously. In today's growing global economy, intermodal facilities have become increasingly popular as a method of increasing efficiency and decreasing costs across the entire spectrum of supply chain operations. In order for a shipment to be considered intermodal it must be handled by more than one mode of transportation such as truck, rail, ship, or plane. The increased focus of efficiency and cost reduction is a product of current shipping trends. The typical freight shipment "traveled nearly 40 percent farther in 2002 than in 1993 (Dipo 2009)." Increased distance traveled for freight implies that the cost associated with shipping has also increased.

Intermodal facilities provide a number of advantages to companies. Intermodal terminals facilitate the transfer of standardized shipping units between modes of transportation. Four primary functions are performed in intermodal facilities: transfer of cargo between modes of transportation, freight assembly in preparation of transfer, freight storage, logistical control and distribution of product flows (Slack 1990). These activities are centralized in order to concentrate critical operations in one location thereby provided opportunities for economies of scale. Strategically placed intermodal facilities within a supply chain provide flexibility to decision makers. These facilities allow operators to select the most efficient method of shipment for each freight container. Increased efficiency implies that less time is wasted on non-value adding activities. Reduced time means money saved while goods are in transit. Additionally, having a shared intermodal facility allows for less capital expenditure on infrastructure; this way companies are moving more freight with fewer assets.

6.2　Containerization

Transportation systems have become dependent on standardized shipping containers to ensure the timely and efficient transfer of containers between shipping modes. Goods that are shipped in standard metal containers of various sizes facilitate mechanized handling for each container. "In this way goods that might have taken days to be unloaded from a ship can now be handled in a matter of minutes (Slack 1998)." Mechanized handling allows the timely transfer between modes. There are many sizes of container available but the most prevalent container size is the 40 foot box, which can carry on average 22 tons of cargo with each haul. The beauty of these containers is that they are airtight, stackable, lockable, and can be outfitted with refrigeration for perishable loads. In the past, pallets were a common management unit in transportation, but their small size and lack of protective frame made transfer between modes of shipment labor intensive. The following chart lists the characteristics of the most popular container sizes. In the United States, a large

Figure 4: Container Size

Container Size and Capacity	
International ISO Sizes (8.6' x 8')	
TEU (20 ft)	
Volume	1,140 cu.ft.
Payload	47,711 lbs
FEU (40 ft)	
Volume	2,390 cu. ft.
Payload	59,040 lbs
Domestic US (9' x 8.25')	
Length	53 ft
Volume	3857 cu. ft.
Payload	67,200 lbs

(element int'l 2011)

amount of domestic carriers utilize the 53 foot containers. Standardized shipping containers offer several advantages to shippers. Shipping containers are capable of shipping many types of products. A company could use the containers for raw material shipments, such as coal or aluminum, then immediately pack the same container with finished goods. Since the containers can ship anything companies do not need to invest as much capital in equipment. The shipment of container transportation also cost as much as twenty times less than regular bulk shipments. Additionally, containers that comply with ISO standards are capable of being handled around the globe by many different companies. As you can imagine, containers that can handled anywhere in the world greatly reduce the time and handling cost associated with any cargo. These containers can be stored anywhere and can be stacked, for storage purposes, multiple units high.

While containerized shipping does have many advantages it does have its downside. The unloading of shipping containers requires a minimum of 12 hectares. This land use requirement could potentially consume a large amount of terminal space, depending on the availability. The mechanized handling equipment mentioned earlier in the chapter, such as gantry cranes, yard equipment, etc. is expensive. These means the companies choosing to ship their products via standardized containers must initially commit to a large capital investment for all primary and secondary equipment. An additional consideration is that as much as 56% of their useful lives (10-15 years) idle or being repositioned. The return on investment for the containers may come quickly but the utilization ratio is generally considered quite low. While storage containers are capable of being stack for storage purposes, the stacking of these containers can be a complex problem. Containers must be stacked using the Last-In-First-Out (LIFO) only. Any other planning method would require a large amount time and money to pull a container from the bottom of a stack. It is true that the use of containerized shipping equipment has its advantages and disadvantages, but the job of the logistics manager is to determine if utilizing standard containers is worth the investment. And in many situations it is.

LECTURE 7

POWER POINT MATERIALS

Transportation Problem

Dr. MD Sarder

Transportation Problem

- **The TP problem has the following characteristics:**
  - m sources and n destinations
  - number of variables is m x n
  - number of constraints is m + n (constraints are for source capacity and destination demand)
  - costs appear only in objective function (objective is to minimize total cost of shipping)
  - Demand and supply must be equal
  - coefficients of decision variables in the constraints are either 0 or 1

Transportation Problem

- Network Representation

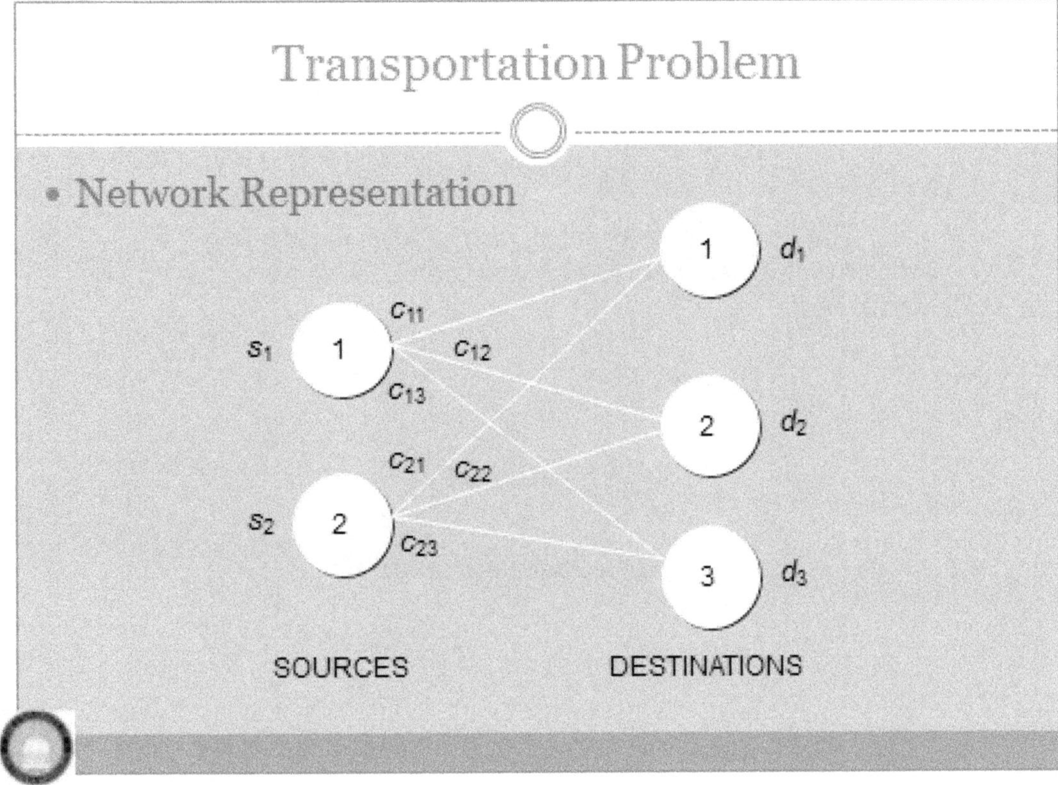

The Relationship of TP to LP

- TP is a special case of LP
- How do we formulate TP as an LP?
  - Let x_{ij} = quantity of product shipped from source i to destination j
  - Let c_{ij} = per unit shipping cost from source i to destination j
  - Let s_i be the row i total supply
  - Let d_j be the column j total demand
- The LP formulation of the TP problem is:

The Relationship of TP to LP

- LP Formulation

  The linear programming formulation in terms of the amounts shipped from the origins to the destinations, x_{ij}, can be written as:

$$\text{Min} \quad \sum_i \sum_j c_{ij} x_{ij}$$

$$\text{s.t.} \quad \sum_j x_{ij} \leq s_i \quad \text{for each origin } i$$

$$\sum_i x_{ij} \geq d_j \quad \text{for each destination } j$$

$$x_{ij} \geq 0 \quad \text{for all } i \text{ and } j$$

Transportation Problem

- LP Formulation Special Cases

The following special-case modifications to the linear programming formulation can be made:

- Minimum shipping guarantees from i to j:
$$x_{ij} \geq L_{ij}$$
- Maximum route capacity from i to j:
$$x_{ij} \leq L_{ij}$$
- Unacceptable routes:

  delete the variable (or put a very high cost, called the Big-M method, on a selected pair of i and j)

Transportation Problem

- To solve the transportation problem by its special purpose algorithm, it is required that the sum of the supplies at the origins equal the sum of the demands at the destinations. If the total supply is greater than the total demand, a dummy destination is added with demand equal to the excess supply, and shipping costs from all origins are zero. Similarly, if total supply is less than total demand, a dummy origin is added.
- When solving a transportation problem by its special purpose algorithm, unacceptable shipping routes are given a cost of $+M$ (a large number).

Transportation Problem

- The transportation problem is solved in two phases:
  - Phase I — Obtaining an initial feasible solution
  - Phase II — Moving toward optimality
- In Phase I, the Minimum-Cost Procedure can be used to establish an initial basic feasible solution without doing numerous iterations of the simplex method.
- In Phase II, the Stepping Stone Method, using the MODI method for evaluating the reduced costs may be used to move from the initial feasible solution to the optimal one.

Transportation Algorithm

- Phase I - Minimum-Cost Method (simple and intuitive)
  - Step 1: Select the cell with the least cost. Assign to this cell the minimum of its remaining row supply or remaining column demand.
  - Step 2: Decrease the row and column availabilities by this amount and remove from consideration all other cells in the row or column with zero availability/demand. (If both are simultaneously reduced to 0, assign an allocation of 0 to any other unoccupied cell in the row or column before deleting both.) GO TO STEP 1.

Example 1: TP

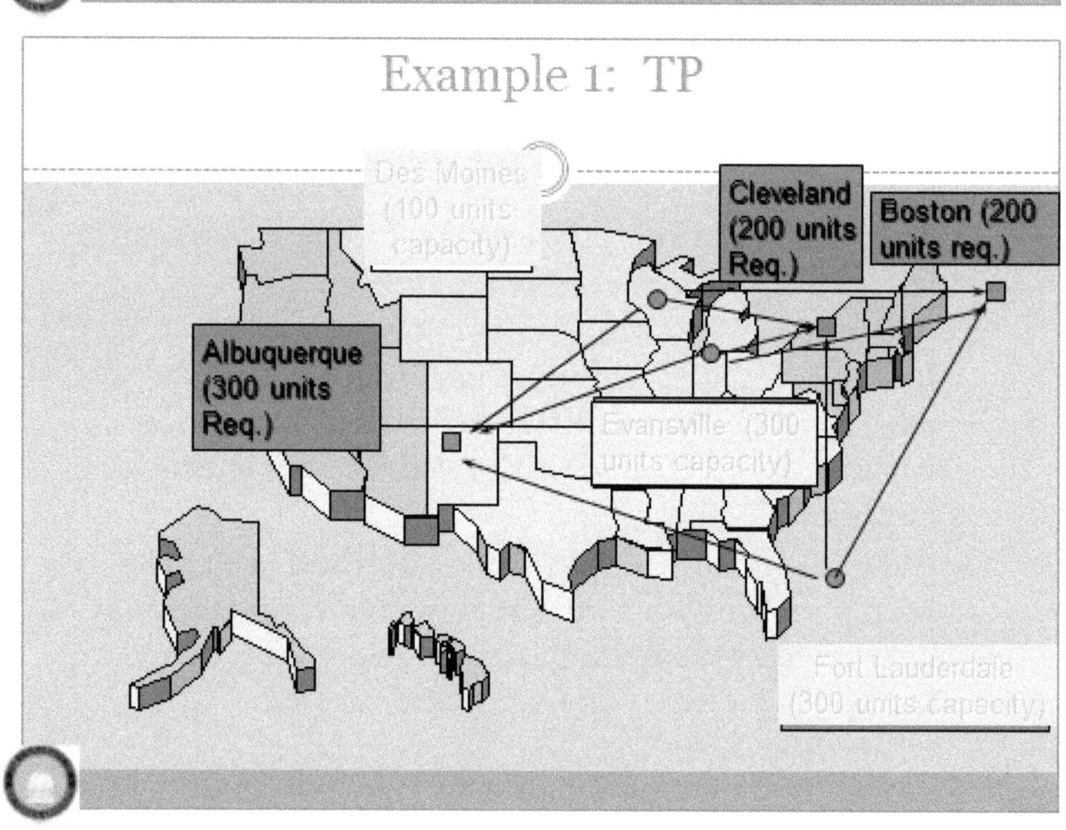

Example 2: TP

	To (Destination)		
From (Sources)	Albuquerque	Boston	Cleveland
Des Moines	$5	$4	$3
Evansville	$8	$4	$3
Fort Lauderdale	$9	$7	$5

Example 2: TP

From	To			Factory capacity
	Albu-querque	Boston	Cleve-land	
Des Moines	5	4	3	100
Evansville	8	4	3	300
Fort Lauderdale	9	7	5	300
Warehouse requirement	300	200	200	700

- ## What is the initial solution?
  - Since we want to minimize the cost, it is reasonable to "be greedy" and start at the cheapest cell ($3/unit) while shipping as much as possible. This method of getting the initial feasible solution is called the least-cost rule
  - Any ties are broken arbitrarily.
    (Des Moines to Cleveland - $3)
    (Evansville to Cleveland - $3)
    Let's start ad cell 1-3 (Des Moines to Cleveland)
  - In making the flow assignments make sure not to exceed the capacities and demands on the margins!

Example 2: TP

| | To | | | Factory |
From	Albu-querque	Boston	Cleve-land	capacity
Des Moines	5	4	**Start** 3 +100	100 ~~~~ 0
Evansville	8	4	3	300
Fort Lauderdale	9	7	5	300
Warehouse requirement	300	200	200 ~~~~ 100	700

Example 2: TP

	To						Factory	
From	Albu-querque		Boston		Cleve-land		capacity	
Des Moines		5		4	÷100	3	100 ~~ 0	
Evansville		8		4	÷100	3	300 ~~ 200	
Fort Lauderdale		9		7		5	300	
Warehouse requirement	300		200		200 ~~ 100 ~~ 0		700	

Example 2: TP

	To						Factory	
From	Albu-querque		Boston		Cleve-land		capacity	
Des Moines		5		4	÷100	3	100 ~~ 0	
Evansville		8	÷200	4	÷100	3	300 ~~ 0 ~~ 200	
Fort Lauderdale		9		7		5	300	
Warehouse requirement	300		200 ~~ 0		200 ~~ 100 ~~ 0		700	

Example 2: TP

From	Albu-querque		Boston		Cleve-land		Factory capacity
					To		
Des Moines		5		4	÷100	3	100̶ / 0
Evansville		8	÷200	4	÷100	3	300̶ / 0 / 2̶0̶0̶
Fort Lauderdale	÷300	9		7		5	300̶ / 0
Warehouse requirement	300̶ / 0		200̶ / 0		200̶ / 1̶0̶0̶ / 0		700

Example 2: TP

- **The initial feasible solution via the least-cost rule is:**
  - Ship 100 units from Des Moines to Cleveland
  - Ship 100 units from Evansville to Cleveland
  - Ship 200 units from Evansville to Boston
  - Ship 300 units from Fort Lauderdale to Albuquerque
- **The total cost of this solution is:**
  - $100(3) + 100(3) + 200(4) + 300(9) = \$4,100$
- **Is the current solution optimal?**
  - May be, may be not, we are not sure. Hence, some more work is needed.

Example 2: TP via LP

- In our example the LP formulation is:

$$\text{Min } z = 5x_{11} + 4x_{12} + 3x_{13} + 8x_{21} + 4x_{22} + 3x_{23} + 9x_{31} + 7x_{32} + 5x_{33}$$

s.t.

$$x_{11} + x_{12} + x_{13} <= 100 \qquad \text{(row 1)}$$
$$x_{21} + x_{22} + x_{23} <= 300 \qquad \text{(row 2)}$$
$$x_{31} + x_{32} + x_{33} <= 300 \qquad \text{(row 3)}$$
$$x_{11} + x_{21} + x_{31} >= 300 \qquad \text{(column 1)}$$
$$x_{12} + x_{22} + x_{32} >= 200 \qquad \text{(column 2)}$$
$$x_{31} + x_{32} + x_{33} >= 200 \qquad \text{(column 3)}$$
$$x_{ij} >= 0 \text{ for } i = 1, 2, 3 \text{ and } j = 1, 2, 3 \text{ (nonnegativity)}$$

Example 2: TP via LP

- The solver formulation and solution is:

				Example 1								
Shipping costs:	5	4	3	8	4	3	9	7	5			
Shipping quantities:	100	0	0	-3.6E-15	200	100	200	0	100		decision cells	
Total shipping cost:	3900										objective function	
s.t.												
Des Moines supply:	1	1	1	0	0	0	0	0	0	100	<=	100
Evansville supply:	0	0	0	1	1	1	0	0	0	300	<=	300
Fort Lauderdale supply:	0	0	0	0	0	0	1	1	1	300	<=	300
Albuquerque demand:	1	0	0	1	0	0	1	0	0	300	>=	300
Boston demand:	0	1	0	0	1	0	0	1	0	200	>=	200
Cleveland demand:	0	0	1	0	0	1	0	0	1	200	>=	200
nonnegativity:	1	0	0	0	0	0	0	0	0	100	>=	0
	0	1	0	0	0	0	0	0	0	0	>=	0
	0	0	1	0	0	0	0	0	0	0	>=	0
	0	0	0	1	0	0	0	0	0	-3.6E-15	>=	0
	0	0	0	0	1	0	0	0	0	200	>=	0
	0	0	0	0	0	1	0	0	0	100	>=	0
	0	0	0	0	0	0	1	0	0	200	>=	0
	0	0	0	0	0	0	0	1	0	0	>=	0
	0	0	0	0	0	0	0	0	1	100	>=	0

Example 2: TP via LP

- The solver answer report is:

Example 2: TP via LP

- The solver sensitivity report is:

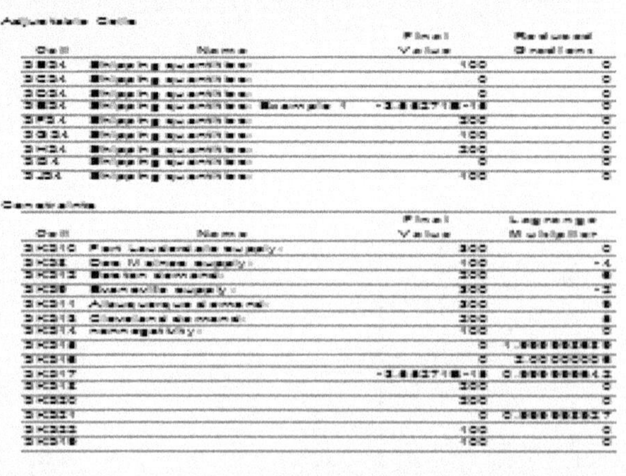

- The solver limits report is:

Microsoft Excel 8.0 Limits Report
Worksheet: [Example 1 - TP via LP.xls]Sheet1
Report Created: 2/6/00 12:12:55 PM

Cell	Target Name	Value
B5	WBMIN	3900

Cell	Adjustable Name	Value	Lower Limit	Target Result	Upper Limit	Target Result
B4	Shipping quantities	100	100	3900	100	3900
C4	Shipping quantities	0	0	3900	0	3900
D4	Shipping quantities	0	0	3900	0	3900
E4	Shipping quantities Example 1	-3.55271E-15	-3.55271E-15	3900	-3.55271E-15	3900
F4	Shipping quantities	200	200	3900	200	3900
G4	Shipping quantities	100	100	3900	100	3900
H4	Shipping quantities	200	200	3900	200	3900
I4	Shipping quantities	0	0	3900	0	3900
J4	Shipping quantities	100	100	3900	100	3900

LECTURE 7
READING MATERIALS

7.0 Introduction

The direction of this paper is to provide an overview of the Transportation Problem methods using linear programming and other approaches. The transportation models are a vital part of the logistics and supply chain organization for decreasing cost and improving service. Therefore, the objective is to find the best cost effective way to transport the goods.
Network points are a representation of how a transportation problem determines how much travel is required from each supply point to each demand point in order to minimize costs.

Deterministic and probabilistic decision models are part of the Decision-making problems. Deterministic models perditions meet expectations based on uncontrollable factor influence and given information. Optimizing system performance in transportation problems necessitates continual improvement to minimize cost and maximize performance. A preferred method is to use a model accurately representing the appropriate aspect of the operation process to describe potential solutions. A mathematical optimization model consists of an objective function and a set of constraints expressed in the form of a system of equations or inequalities. Optimization solution methodologies based on simultaneous thinking result in optimal solution. The systematic approach is an optimization solution algorithm [10].

7.1 Transportation problem defined

One of the most important and successful applications of quantitative analysis to solving business problems has been in the physical distribution of products, commonly referred to as transportation problems. Basically, the purpose is to minimize the cost of shipping goods from one location to another so that the needs of each arrival area are met and every shipping location operates within its capacity [5].
A transportation problem (TP) maybe defined as any task concerned with the distribution of products from any supply source to any demand destinations at the lowest possible distribution cost. Each supply source has a pre-determined supply capacity and each demand destination has a pre-determined level of demand that has to be satisfied. The objective of a transportation problem is to ensure that the correct amount of products are shipped between each source and destination while minimizing shipping cost and meeting demand and supply constraints. Figure I is an Illustration of supply to demand shipment [11].

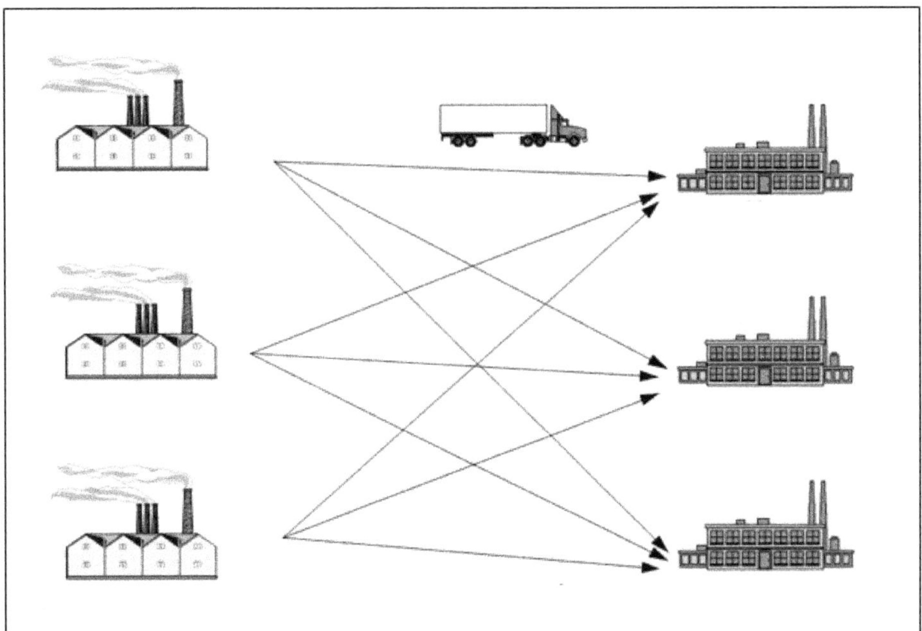

Figure I Illustration of supply to demand shipment

The transportation problem solution methods key purpose is to reduce the cost of the transportation time. The majority of the presently used methods for solving transportation problems are trying to reach the optimal solution.

The sole objective is to ensure that the right quantities of products arrive in the right place, at the right time and in the best condition, that will make the greatest contribution to the company.

The transportation problem can be balanced or unbalance. A balanced transportation problem is when the total amount demanded at all destinations is equal to the total supply available at all sources [9]. An unbalanced transportation problem is when the total demanded at all destinations is not equal to the total supply available at all sources. Even if the transportation problem is unbalanced, it is possible to balance the problem. To balance an unbalanced transportation problem you need to add a dummy line. If your total supply is less than your total demand, you add a dummy source and let the supply equal the difference. The shipping cost in a dummy destination or shipping cost from a dummy source is always zero.

7.2 Solving transportation problems

In order to solve a transportation problem you must first understand the necessary steps it takes to solve the problem. There are four common steps used in solving a transportation problem they are as followings:

- Define the problem and prepare the transportation tableau
- Obtain an initial feasible solution
- Identify the optimal solution
- Understand special situations

7.2.1 Define the problem and prepare the transportation tableau

The transportation tableau clearly articulates the supply and demand constraints and the shipping cost between each demand and supply point.
Below is an example of a transportation tableau.

Table I Example Transportation cost Matrix

From Source	To Destination				Factory Capacity
	D_1	D_2	D_3	D_4	
S_1					
S_2					
S_3					
S_4					
Warehouse Requirements					

7.2.2 Obtain an initial feasible solution method

There are various initial basic feasible solution used to solve a transportation problem. Here are a few of commonly used methods.

I.   North-West Corner Method
2.   MODI Method
3.   Floating Point Method
4.   Fourier Method
5.   Blocking Method
6.   Simplex Method
7.   Excel Solver
8.   Lindo Method

7.2.2.a North-West Corner Method for Solving Transportation

The North-west corner method even though it may be simple and easy use it often produces an insufficient solution which can be a considered a disadvantage. Steps and example will be shown in section 4.4 [5].

7.2.2.b Modi Method For Solving Transportation

The MODI method uses the following steps [4].

Step I: Add u_i col and v_j row: Add a column on the right hand side of the TP table and title it u_i. Also add a row at the bottom of the TP table and title it v_j.

Step 2: This step has four parts.

i) Assign value to $u_i=0$ To any of the variables u_i or v_j, assign any arbitrary value. Generally the variable in the first row i.e. u_I is assigned the value equal to zero.

ii) Determine values of the v_j in the first row using the value of $u_I = 0$ and the c_{ij} values of the occupied cells in the first row by applying the formula $u_i + v_j = c_{ij}$

iii) Determine u_i and v_j values for other rows and columns with the help of the formula $u_i + v_j = c_{ij}$ using the u_i and v_j values already obtained in steps i), ii) above and c_{ij} values of each of the occupied cells one by one.

iv) Check the solution for degeneracy. If the solution is degenerate [ie no. of occupied cells is less than $(m+n-I)$], then this method will not be applicable.

Step 3: Calculate the net opportunity cost for each of the unoccupied cells using the formula $\delta_{ij} = (u_i + v_j)$ - c_{ij}. If all unoccupied cells have negative δ_{ij} value, then, the solution is optimal. Multiple optimality: If, however, one or more unoccupied cells have δ_{ij} value equal to zero, then the solution is optimal but not unique. Non optimal solution: If one or more unoccupied cells have positive δ_{ij} value, then the solution is not optimal. Largest positive d_j value: The unoccupied cell with the largest positive δ_{ij} value is identified.

Step 4: A closed loop is traced for the unoccupied cell with the largest δ_{ij} value. Appropriate quantity is shifted to the unoccupied cell and also from and to the other cells in the loop so that the   transportation cost comes down.

Step 5: The resulting solution is once again tested for optimality. If it is not optimal, then the steps from I to 4 are repeated, till an optimal solution is obtained.

7.2.2.c Floating Point Method for Solving Transportation

In the Floating Point method, is used for finding an optimal solution to a transportation problem with additional constraints [19]. The optimal solution obtained by floating point method is not necessarily to have $(m + n - I)$ allotted entries which is the main advantage of this method. The floating point method provides an appropriate solution which helps the decision makers to analyze economic activities and to arrive at the correct managerial decisions.

The following are steps for the Floating Point Method [3].

Step I: Identify all impossible allotment cells using additional constraints and remove those cells from the given TP.

Step 2: Obtain an optimal solution to the reduced TP obtained from Step I. Without considering the additional constraints using the zero point method [9].

Step 3: Check that all additional constraints are satisfied for the optimal solution obtained in the Step 2. If so, go to the Step 7. If not, go to the Step 4.

Step 4: Identify the column(s) which is (are) not satisfied the additional constraints.

Step 5: Select a loop in a column which is not satisfied the additional constraint and flow the minimum quantity θ according to the Theorem I.. Then, find a new solution to the transportation problem   with additional constraints.

Step 6: Check that the solution obtain from the Step 5 satisfies all additional constraints. If so, go to the Step 7. If not, go to the Step 4.

Step 7: The current solution is an optimal solution to the TP with all additional constraints by the Theorem I..

7.2.2.d Fourier Method for Solving Transportation

In the Fourier transportation is used for finding an optimal solution of transportation problems with mixed constraints. This method is very easy to understand and apply which can serve managers by providing the optimal solution to a variety of distribution problems with mixed constraints.

The Fourier transportation method steps are [2]:

Step 1: Write the given transportation problem with mixed constraints in the form of a pure integer linear programming problem.

Step 2: Convert the pure integer linear programming problem obtained from the Step 1.into a maximization problem.

Step 3: Write the maximization problem obtained from the Step 2 Having only one type of the inequality ' $\leq$ ' by eliminating one variable in an equality constraint in the given problem.

Step 4: Write the equivalent pure integer linear programming problem to the modified maximization problem obtained from the Step 3 and then, construct the Fourier elimination table for the equivalent problem.

Step 5: Select and remove a variable from the Fourier elimination table by the Modified Fourier's elimination method [6].

Step 6: Form a reduced Fourier elimination table from the Fourier elimination table after deleting the true statement(s) and redundant constraints (if any) and also, using the Theorem 2.

Step 7: Repeat the Step 5 to the Step 6 until all variables of x_{ij} ' s are eliminated except the objective function variable w .

Step 8: Find the least upper bound of all maximum possible values of w. The optimum solution w is d o . The values of all x_{ij} ' s are computed using backward substitution method and basic algebraic method. Say x_{ijo} , i = 1,2,...,m and j = 1,2,...,n.

Step 9: The optimal solution for the transportation problem with mixed constraints is $x_{ij} = x_{ijo}$, i = 1,2,...,m and j = 1,2,..., n and the minimum value of z , z o = d o .

7.2.2.e The Blocking Method For Solving Transportation

The Blocking method involves defining a solution for using the least-cost method.

The steps are as follows [6]:

Step 1: Find the maximum of the minimum of each row and column of the given transportation table. Say, T.

Step 2: Construct a reduced transportation table from the given table by blocking all cells having time more than T.

Step 3: Check if each column demand is less than to the sum of the supplies in the reduced transportation problem obtained from the Step 2.. Also, check if each row supply is less than to sum of the column demands in the reduced transportation problem obtained from the Step 2.. If so, go to Step 6. (Such reduced transportation table is called the active transportation table). If not, go to Step 4.

Step 4: Find a time which is immediately next to the time T. Say U.

Step 5: Construct a reduced transportation table from the given   transportation table by blocking all cells having time more than U and then, go to the Step 3.

Step 6: Do allocation according to the following rules:

(a) Allot the maximum possible to a cell which is only one cell in the row/column. Then, modify the active transportation table and then, repeat the process till it is possible or all allocations are completed. (b) If (a) is not possible, select a row / a column having minimum number of unblocked cell and allot maximum possible to a cell which helps to reduce the large supply and/ or large demand of the cell.

Step 7: This allotment yields a solution to the given bottleneck transportation problem. Now, we prove the solution to a BTP obtained by the blocking method is an optimal solution to the BTP. Remark 2: The time or / and cost budgetary transportation problem by the blocking

7.3 Methods for solving transportation problem

Network models:

Efficiently and effectively solve decision problems

Decision problems modeled using network optimization

Optimizes transportation or flow of supplies

Decision variables are expected to be continuous

Network optimization problems model:

Transportation and min cost flow problems

Use transportation network for roads, river channels, or flight patterns

Supply connectivity between networks

Large numbers of possible routes use special solution algorithms

Transportation Problem:

Objective is to determine minimum transportation costs

Optimizes between warehouse and destination

Goal is effective transport of goods

Uses designated supply source and destination

The objective of a transportation problem is to schedule shipments from sources to destinations so that total transportation costs are minimized.   Typically, there is a source capacity and destination requirement as shown in the following example of Table 2 below.

Table [2] Source and destination capacity and requirements

SOURCES			W/H DESTINATIONS	
CAPACITIES		ROUTES		REQUIREMENTS
1000	DES		ALBU	1000
2000	EVA		BOS	200
700	FT.LAU		CLEV	2500

LP can solve the following kind of problem, but the transportation method uses simple computation with more proficient special-purpose algorithms. "To begin the analysis of a transportation problem, management must determine time costs of shipping from each source to each destination." The Chocolate Delights Corporation, a distributor of candy and sweets presents data in tabular form," as seen in Table 3 below.

Table [3] Source and destination tabular form data

From Supply	To Demand			
	ALBU	BOS	CLEV	Factory Capacity
DES	1	8	5	1000
EVA	7	6	4	2000
FT.LAU	9	4	4	700
Warehouse Requirements	1000	200	2500	**3700**

Next construct a transportation table to summarize data and track the systematic approach of algorithm calculations.

Table [4] Source and destination tabular form data

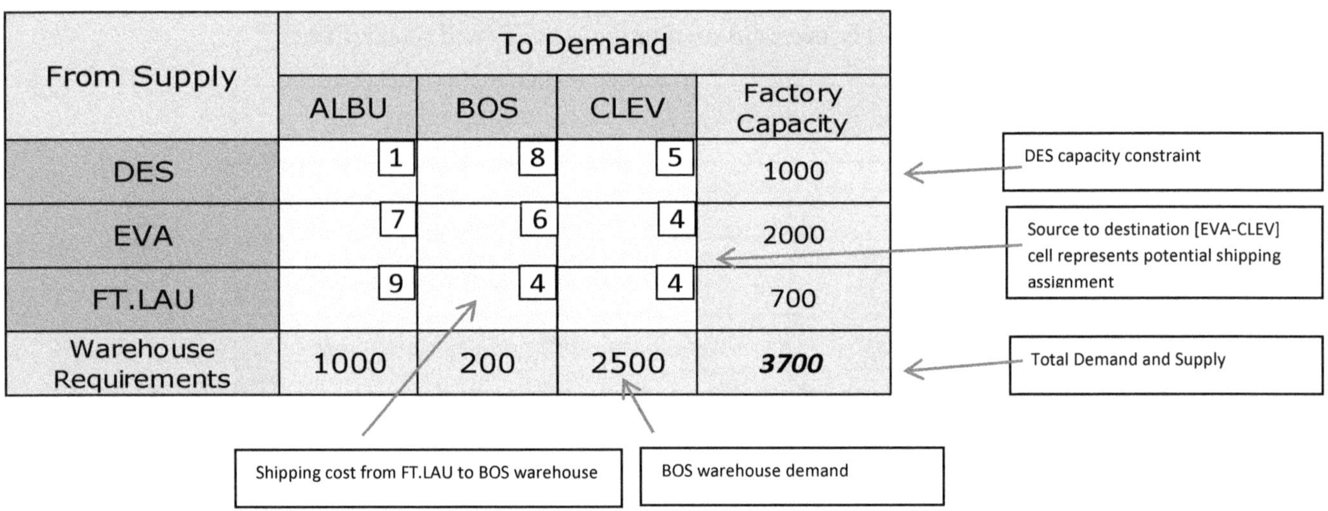

From Supply	To Demand			
	ALBU	BOS	CLEV	Factory Capacity
DES	1	8	5	1000
EVA	7	6	4	2000
FT.LAU	9	4	4	700
Warehouse Requirements	1000	200	2500	**3700**

THE NORTHWEST CORNER method develops an initial solution

Using a systematic procedure

Arranging data in tabular form

Establishing an initial feasible solution

Starting in upper left hand cell (northwest corner) of table

Allocating units to shipping routes as follows:

Exhaust supply (factory supply) at each row before moving to next row

Exhaust (warehouse) supplies of column before moving to next column

Check that supply and demands are met

EXAMPLE I.I

We can use the northwest corner rule to find an initial feasible solution to the Chocolate Delights Corporation problem shown in Table 5 below.

It takes four steps in this example to make the initial distribution

Assign 1000 units from Des to ALBU (exhausting ALBU's demand and DES supply).

Assign 2000 units from EVA to CLEV (exhausting EVA's supply).

Assign 200 units from FT.LAU to BOS (exhausting BOS demand).

Assign 500 units from FT.LAU to CLEV (exhausting CLEV's demand and FT.LAU's supply).

Table [5] Source and destination tabular form data

From Supply	To Demand			
	ALBU	BOS	CLEV	Factory Capacity
DES	1000 [1]	[8]	[5]	1000
EVA	[7]	[6]	2000 [4]	2000
FT.LAU	[9]	200 [4]	500 [4]	700
Warehouse Requirements	1000	200	2500	*3700*

Effortlessly compute the cost of shipping in Table 6 below

Table [6] Source and destination tabular form data

	ROUTE			
FROM	TO	UNITS SHIPPED	UNIT COST	TOTAL COST
DES	ALBU	1000	$1	1000
EVA	CLEV	2000	$4	8000
FT.LAU	BOS	200	$4	800
FT.LAU	CLEV	500	$4	2000
				$11,800

The solution given here is feasible since demand-and-supply constraints are all satisfied.

It would be very lucky if this solution yielded the minimal transportation cost for the problem, however

likely that one of the iterative procedures designed to help reach an optimal solution shall have to be employed.

First, try to use the method to solve Problem 13.1.

Demand Not Equal To Supply

A situation occurring quite frequently in real-world problems is the case where total demand is not equal to total supply.

These unbalanced problems can be handled easily by the solution procedures discussed above if we first in introduce dummy sources or dummy destinations.

In the event that total supply is greater than total demand, a dummy destination, with demand exactly equal to the surplus, is created.

If total demand is greater than total supply, we introduce a dummy source (factory) with a supply equal to the excess of demand over supply.

In either ease, cost coefficients of zero are assigned to each dummy location.

EXAMPLE 1.2

Chocolate Delights Increases the rate of distribution of chocolate in its FT.LAU warehouse to 1000. To reformulate this unbalanced problem, we refer back to the data presented in Example 1.1. The northwest corner rule is used to find the initial feasible solution in Table 7 below.

Table [7] Source and destination tabular form data

From Supply	To Demand				
	ALBU	BOS	CLEV	DUMMY	Factory Capacity
DES	1000 |1|	|8|	|5|	|0|	1000
EVA	|7|	|6|	1500 |6|	500 |0|	2000
FT.LAU	|9|	200 |4|	1000 |4|	|0|	1200
Warehouse Requirements	1000	200	2500	*500*	*4200*

Total cost = $(1000)(\$1) + (200)(\$4) + (1500)(\$6) + 1000(\$4) + 500(\$0) = \$14{,}800$

The Modi Method

The MODI (modified distribution) method allows us to compute indices for each unused square without drawing all the closed paths.

Because of this, it can often provide considerable time savings over the stepping-stone method for solving transportation problems.

In applying the MODI method, we begin with an initial solution obtained by using the northwest corner rule.

But now, we must compute a value for each row (call the values R1, R2, R3 if there are 3 rows) and for each column (K1, K2, K3) in the transportation table.

In general, we let

R_i = value assigned to row i.

K_j = value assigned to column j.

C_{ij} = cost in square ij (cost from source i to destination j).

The MODI method then requires three steps:

To compute the values for each row and column, set $R_i + K_j = C_{ij}$ but only for those squares that are currently used or occupied.

After all equations have been written, set $RI = 0$.

Solve the system of equations for all R and K values.

Compute the improvement index for each unused square by the   formula: $C_{ij} - R_i - K_j$.

Select the largest negative index and proceed to solve the problem as we did using the stepping-stone method.

EXAMPLE I.3

Given the initial solution to the Chocolate Delights Corporation problem (from Example I.I), we can use the MODI method to calculate an improvement index for each unused square. The initial transportation table is repeated below.

		K_1	K_2	K_3	
	From Supply	**To Demand**			
		ALBU	**BOS**	**CLEV**	**Factory Capacity**
R_1	**DES**	1000 [1]	[8]	[5]	1000
R_2	**EVA**	[7]	[6]	2000 [4]	2000
R_3	**FT.LAU**	[9]	200 [4]	500 [4]	700
	Warehouse Requirements	1000	200	2500	**3700**

Table [7] Source and destination tabular form data

We first set up an equation for each occupied square:

RI + KI = I

R2 + K3 = 4

R3 + K2 = 4

R3 + K3 = 4

Letting RI = 0, we can easily solve, step by step, for KI, R2, K2, R3 and K3

0 + KI = I → KI = I

R2 + I = 7 → R2 = 6

6 + K3 = 4 → K3 = -2

R3 -2 = 4 → R3 = 6

6 + K2 = 4 → K2 = -2

7.4 linear programming

Linear programming minimizes shipping cost while meeting demand. Figure 2 [I] shows a network representation and the information is needed to set up such a model.

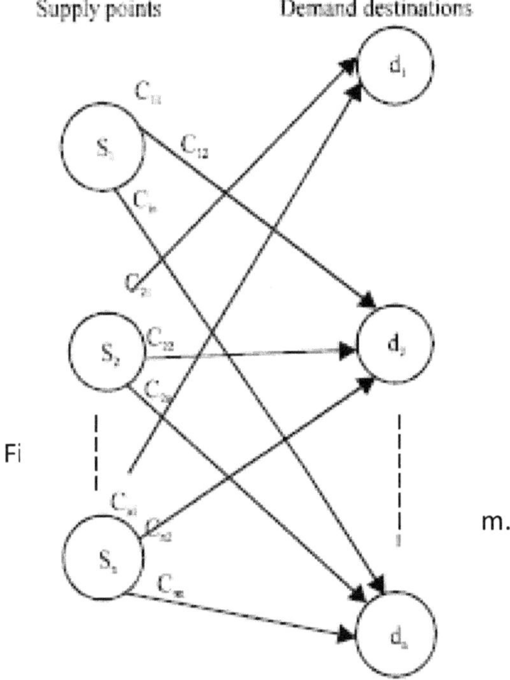

- A set of n supply points from which product are shipped. Supply point I can supply at most s_i units.

- A set of m demand destinations to which the product is shipped. Demand destination j must receive at least d_j units of the shipped product

- Each plant reduced at supply point I and shipped to demand destination j incur a variable cost of C_{ij}

149

The TP problem has the following characteristics:

- m sources and n destinations
- number of variables is m x n
- number of constraints is m + n (constraints are for source
- capacity and destination demand)
- costs appear only in objective function (objective is to minimize total cost of shipping)
- Demand and supply must be equal
- coefficients of decision variables in the constraints are either 0 or I

7.5 solving transportation problems

Explain TP

Solved using simplex method

Solved using excel solver method

Solved using Northwest corner method

Linear programming is a resources allocating method, which finds the optimal solution. It is one of the most widely used operations research tools that are used for decision making in almost all manufacturing industries and also in service organizations. [7]

Linear programming, programming is a mathematical programming tool that allocates resources— labor, materials, machines, capital—in the best possible (optimal) way so that costs are minimized or profits are maximized. The criterion for selecting the best values of the decision variables can be described by a linear function of these variables; that is, a mathematical function involving only the first powers of the variables with no cross-products. For example, $aX2$ and $bX3$ are valid decision variables, while $aX22$, $bX33$, and $(4XI * 2XI)$ are not. The entire problem can be expressed in terms of straight lines, planes, or analogous geometrical figures.

The terms used in LP are described below:

Decision variables: In LP, these resources are known as decision variables.

Objective function: The criterion for selecting the best values of the decision variables (e.g., to maximize profits or minimize costs) is known as the objective function.

Constraint: Limitations on resource availability form what is known as a constraint set.

7.5.a   Using Lindo to solve transportation problems

LP problems can be solved by a software named LINDO. An example is described below about how to solve transportation problem using this software:

A Sawmill company let us assume the name is XYZ management wants to improve their delivery schedule to keep a steady, adequate flow of logs to his sawmills to capitalize on the good lumber market. Another objective is to reduce the transportation cost. The harvesting group plans to move to three new logging sites. The distance from each site to each sawmill is in Table 1. The average haul cost is $2 per mile for both loaded and empty trucks. The logging supervisor estimated the number of truckloads of logs coming off each harvest site daily. The number of truckloads varies because terrain and cutting patterns are unique for each site. Finally, the sawmill managers have estimated the truckloads of logs their mills need each day [12]. All these estimates are in Table 8.

Table 8: Supply and demand of saw logs for the XYZ Sawmill Company

Logging site	Distance to mill (miles)			Maximum truckloads/day Per logging site
	Mill A	Mill B	Mill C	
1	8	15	50	20
2	10	17	20	30
3	30	26	15	45
Mill demand (truckloads/day)	30	35	30	

The next step is to determine costs to haul from each site to each mill (Table 9).

Table 9: Round trip transportation costs for XYZ Sawmill Company

Logging site	Mill A	Mill B	Mill C
1	(8 miles×2) × ($2 per mile) =$32	$ 60	$ 200
2	(10 miles ×2) × ($2 per mile)= $40	68	80
3	$120	104	60

The management of XYZ wants to minimize hauling cost and meet each of the sawmills' daily demand while not exceeding the maximum number of truckloads from each site. Therefore we can set the LP problem up as a cost minimization. We can formulate the problem as:

Let X_{ij} = Haul costs from Site i to Mill j

i = 1, 2, 3 (logging sites) j = 1, 2, 3 (sawmills)

Objective function:

MIN $32X_{11} + 40X_{21} + 120X_{31} + 60X_{12} + 68X_{22} + 104X_{32} + 200X_{13} + 80X_{23} + 60X_{33}$

Subject to:

$X_{11} + X_{21} + X_{31} > 30$ Truckloads to Mill A

$X_{12} + X_{22} + X_{32} > 35$ Truckloads to Mill B

$X_{13} + X_{23} + X_{33} > 30$ Truckloads to Mill C

$X_{11} + X_{12} + X_{13} < 20$ Truckloads from Site 1

$X_{21} + X_{22} + X_{23} < 30$ Truckloads from Site 2

$X_{31} + X_{32} + X_{33} < 45$ Truckloads from Site 3

$X_{11}, X_{21}, X_{31}, X_{12}, X_{22}, X_{32}, X_{13}, X_{23}, X_{33} > 0$

For the computer solution: in the edit box of LINDO, type in the objective function, then "Subject to," then list the constraints. Note, the non-negativity constraint does not have to be typed in because LINDO knows that all LP problems have this constraint. Now the software will solve the problem. It will add slack, surplus, and artificial variables when necessary.

The LINDO (partial) output for the XYZ Sawmill Company transportation problem:

LP OPTIMUM FOUND AT STEP 3

OBJECTIVE FUNCTION VALUE

5760.000

VARIABLE VALUE REDUCED COST

X_{11} 20.000000 0.000000

X_{21} 10.000000 0.000000

X_{31} 0.000000 44.000000

X_{12} 0.000000 0.000000

X_{22} 20.000000 0.000000

X_{32} 15.000000 0.000000

X_{13} 0.000000 184.000000

X_{23} 0.000000 56.000000

X_{33} 30.000000 0.000000

ROW SLACK OR SURPLUS DUAL PRICES

2) 0.000000 −76.000000

3) 0.000000 −104.000000

4) 0.000000 −60.000000

5) 0.000000 44.000000

6) 0.000000 36.000000

7) 0.000000 0.000000

NO. ITERATIONS = 3

Interpretation of the LINDO output

It took three iterations, or pivots, to find the optimal solution of $5,760. (To solve this small LP by hand would have required computations for at least three simplex tableaus.)

The $5,760 represents the minimum daily haul costs for the XYZ Sawmill Company from the three logging sites to the three sawmills. We can use the values in the VALUE column to assign values to our variables and determine the log truck haul schedules.

For variable X_{11}, which is Site 1 to Mill A, the VALUE— number of truckloads per day—is 20 (Table 10).

For variable X_{21}, which is Site 2 to Mill A, the VALUE is 10; and from X_{31}, Site 3 to Mill A, the VALUE is zero, so no loads will be hauled from Site 3 to Mill A. For X_{22}, Site 2 to Mill B, the daily number of truckloads will be 20, and so on.

Table 10: XYZ Sawmill Company log truck haul schedule

Logging site	Mill	Truckloads per day	Cost per load	Total cost
1	A	20	$ 32	$ 640
1	B	0	60	0
1	C	0	200	0
2	A	10	40	400
2	B	20	68	1360
2	C	0	80	0
3	A	0	12	0
3	B	15	104	1560
3	C	30	60	1800

7.5.b Using excel solver to solve transportation problems

Transportation problems can be solved by using excel solver data analysis tool. In this part we are discussing a management decision making problem that is solved as transportation problem. In this problem two telecom companies that want to be unified which have five factories currently. The decision factor is, if they merged then which factories will be required to run their operation in order to serve the six markets. Management can select plant location selection with capacity constraint for single sourcing. In this option one market will be supported by one factory, which is referred as single source. This option lowers the complexity of coordinating the network and requires less flexibility from each facility [13].

Mathematical model:

Notations:

n = number of factory locations

m = number of markets or demand points

D_j = annual demand from market j

K_i = capacity of factory i

c_{ij} = total cost of producing and shipping one unit from factory i to market j

total cost = production cost + transportation cost + inventory cost

y_i = I if factory i is open, 0 otherwise

x_{ij} = quantity shipped from factory I to market j

Objective function:

$$Min \sum_{i=1}^{n} f_i y_i + \sum_{i=1}^{n} \sum_{j=1}^{m} c_{ij} x_{ij}$$

Constraints:

All decision variables ≥ 0

$$K_i y_i - \sum_{j=1}^{m} x_{ij} \geq 0 \ for \ i = 1, \ldots \ldots, 5$$

$$D_j - \sum_{i=1}^{n} x_{ij} = 0 \ \ for \ j = 1, \ldots, 6$$

All location variables are binary, that is : y_i = 0 or I

The following table represents capacity and demand data for all factories and market respectively. Total cost that is cost of production, transportation and inventory cost are also listed here.

Table II: Total cost and demand and capacity data for the unified telecom company

Supply City	Columbus	Hattiesburg	Hammond	Gulfport	Atlanta	Albany	Fixed Cost ($)	Capacity
	Demand City — Production and Transportation Cost per 1000 Units							
Huntsville	1,675	400	685	1,630	1,160	2,800	8,000	30
Beaumont	1,460	1,940	970	100	495	1,200	4,000	25
Henderson	1,925	2,400	1,425	500	950	800	5,000	27
Shreveport	380	1,355	543	1,045	665	2,321	4,100	26
Natchez	922	1,646	700	508	311	1,797	2,000	15
Demand	15	10	20	5	10	11		

Figure: Excel solver parameter entry interface

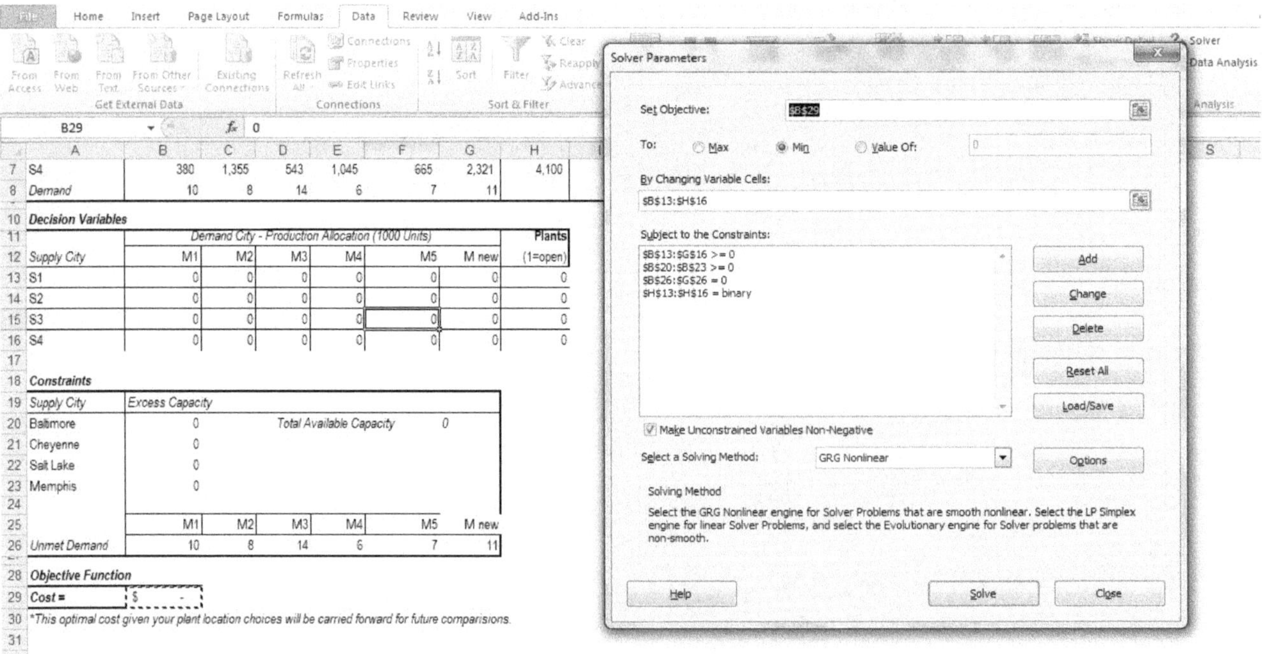

Analyzing the objective function and costs, it is found that the unified company can close one factory (P3) to efficiently serve existing six markets.

Table 12: Optimal Network design for the unified company

| Supply City | Demand City - Production Allocation (1000 Units) | | | | | | Plants |
	Columbus	Hattiesburg	Hammond	Gulfport	Atlanta	Albany	(1=open)
Huntsville	0	10	9	0	0	0	1
Beaumont	0	0	0	5	0	11	1
Henderson	0	0	0	0	0	0	0
Shreveport	15	0	11	0	0	0	1
Natchez	0	0	0	0	10	0	1

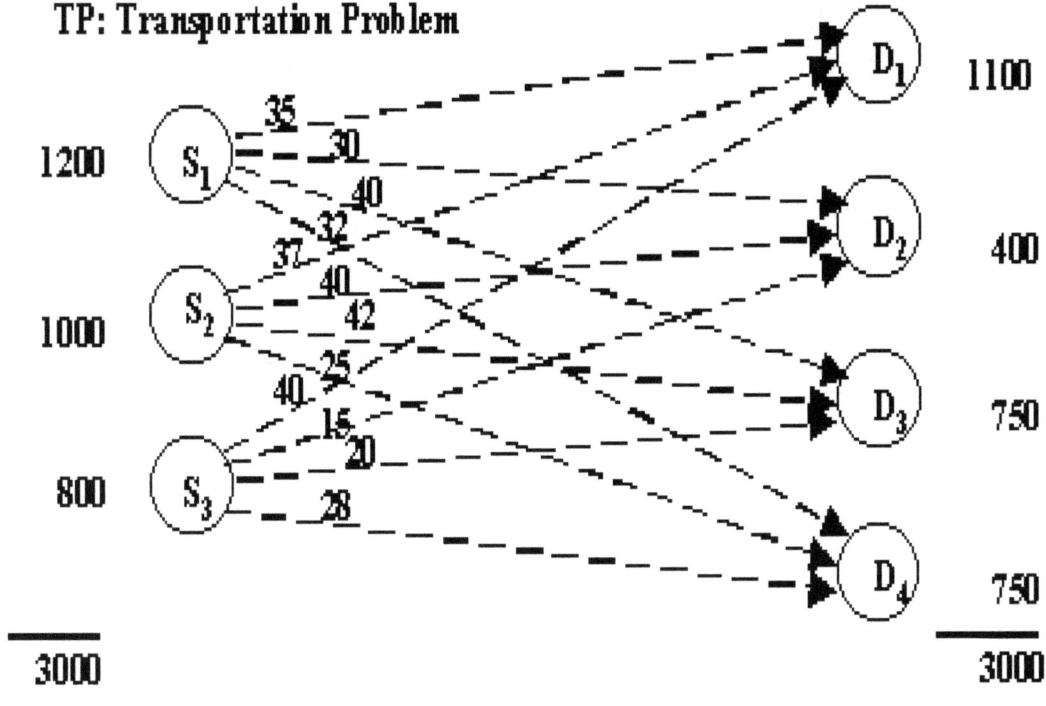

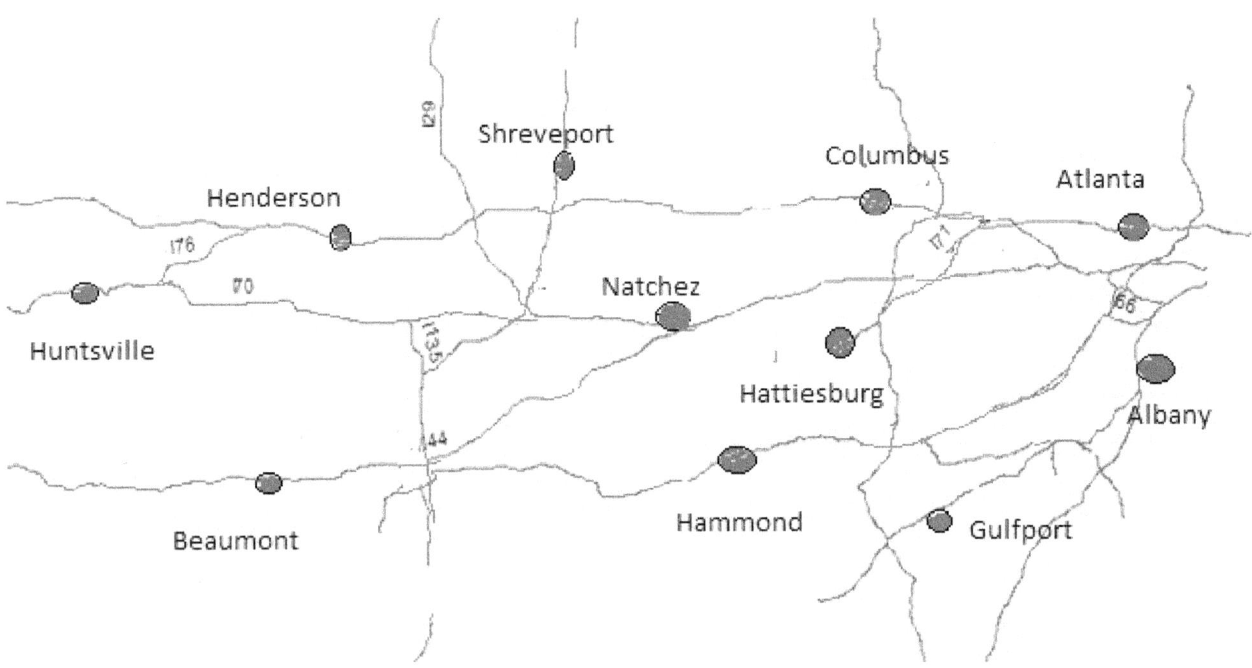

If the two companies run separately with their products it will cost \$64,202 while after unifying they can close one factory and the total cost is \$47,401 which is much more less than previous.

7.5.c Using simplex technique to solve transportation problems

For this type of problem, all units available must be supplied. From the above problem, we see this in fact occurs: the sawmills use all 95 truckloads available. But isn't this unreasonable? How likely it is that supply always will equal demand? For practical purposes, it doesn't matter. As long as supply is adequate to meet demand, we can ignore the surplus and treat the total supply as equal to the total requirement [12].

Also, the number of constraints must equal the number of rows and number of columns when we set up our transportation problem.

We have met this requirement also: three rows plus three columns equals six constraints, as shown above when we set up the problem to enter into the computer.

Another requirement is that the number of routes should equal the number of sources (sites), S, plus the number of destinations (mills), D, minus one, or

$$R = S + D - 1$$

For problems in which $S + D - I$ does not equal the number of routes, either excess routes or degeneracy (more than one exiting cell) can occur. We will talk about these problems on page 14.

Let's set up our problem in a table format (Table 4). There are a number of methods of solving transportation problems. We will look at a common one, the northwest corner method (Table 5a).

In northwest corner method we start work in the northwest corner, or the upper left cell, Site 1 Mill A. Make an allocation to this cell that will use either all the demand for that row or all the supply for that column, whichever is smaller. We see that Site 1's supply is smaller than Mill A's demand, so place a 20 in cell Site 1 Mill A. This eliminates row Site 1 from further consideration because we used all its supply. The next step is to move vertically to the next cell, Site 2 Mill A, and use the same criteria as before.

Now the smaller value is column Mill A (versus 30 in row Site 2); there are only 10 truckloads demanded because 20 have been supplied by Site 1. So, move 10 into this cell. Now we can eliminate column Mill A from further consideration. Move to the next cell, Site 2 Mill B (Table 5b). Move the smaller of the supply or demand values into this cell. Site 2 supply is 20 (remember, we used 10 of the 30 in Site 2 Mill A); Mill B demand is 35, so move 20 into this cell. We can eliminate row Site 2 from further consideration since all its supply (30) is used.

Move to cell Site 3 Mill B. The smaller of the margin values is 15. (Mill B demand is 15 because 20 of 35 truckloads have been supplied by Site 2.) Place 15 in this cell and eliminate column Mill B from further consideration since its demand has been met [12].

Finally, move to cell Site 3 Mill C. Margin values are tied at 30 (vmillC $= 30$ and usite $3 = 30$ because we previously allocated 15 of the 45 available to Site 3 Mill B), so place 30 in this cell. We check to see whether we've met the requirement that the number of routes equals the number of sites plus the number of mills minus 1. In fact, we have met the requirement $(3 + 3 - 1 = 5)$ and we have five routes.

Our initial basic feasible solution is:

$32X_{11} + 40X_{21} + 68X_{22} + 104X_{32} + 60X_{33}$

$\$32(20) + \$40(10) + \$68(20) + \$104(15) + \$60(30) = \$5,760$

However, this may not be our optimal solution. (Actually, we know it is optimal because of our LINDO solution, above.) Let's test to see whether the current tableau represents the optimal solution. We can do this because of duality theory.

We can introduce two quantities, u_i and v_j, where u_i is the dual variable associated with row i and v_j is the dual variable associated with column j. From duality theory:

$X_{ij} = u_i + v_j$ (Eq. I)

We can compute all u_i and v_j values from the initial tableau using

Eq. I.

$X_{II} = u_I + v_2 = 32$

$X_{2I} = u_2 + v_I = 40$

$X_{22} = u_2 + v_2$

$= 68$

$X_{32} = u_3 + v_2 = 104$

$X_{33} = u_3 + v_3 = 60$

Since there are M + N unknowns and M + N – I equation, we can arbitrarily assign a value to one of the unknowns. A common method is to choose the row with the largest number of allocations (this is the number of cells where we have designated truckloads of logs). Row Site 2 and row Site 3 both have two allocations [12]. We arbitrarily choose row Site 2 and set $u_2 = 0$. Using substitutions, we calculate:

$u_2 = 0$

$X_{2I} = u_2 + v_I = 40\ 0 + v_I = 40\ v_I$

$= 40$

$X_{22} = u_2 + v_2 = 68\ 0 + v_2 = 68\ v_2 = 68$

$X_{II} = u_I + v_I = 32\ u_I + 40 = 32\ u_I = -8$

$X_{32} = u_3 + v_2 = 104\ u_3 + 68 = 104\ u_3 = 36$

$X_{33} = u_3 + v_3 = 60\ 36 + v_3 = 30\ v_3 = -6$

Arrange the u_i and v_j values around the table margin (Table 5c).

Here's how to recognize whether this tableau represents the optimal solution: for every non-basic variable (those cells without any allocations), $X_{ij} - u_i - v_j > 0$ (Eq. 2). If Eq. 2 is true in every case, then the current tableau represents the optimal solution; if it is false in any one case, there is a better solution.

For cell Site 1 Mill B: $60 - (-8) - 68 > 0$ is true.

For cell Site 1 Mill C: $200 - (-8) - (-6) > 0$ is true.

For cell Site 2 Mill C: $80 - 0 - (-6) > 0$ is true.

For cell Site 3 Mill A: $120 - 36 - 40 > 0$ is true.

We know this represents the optimal solution and that $5,760 is our lowest cost to ship logs.

Table 4.—Hauling costs, log truckloads available, and log truckloads demanded.

Logging sites	Mill A	Mill B	Mill C	Available truckloads
Site 1	$ 32	$ 60	$ 200	20
Site 2	40	68	80	30
Site 3	120	104	60	45
Sawmill demand:	30	35	30	

Table 5a.—XYZ Sawmill Company transportation problem.

	Mill A	Mill B	Mill C	Supply	u_i
Site 1	32 20	60	200	20	
Site 2	40 10	68	80	30	
Site 3	120	104	60	45	
Demand	30	35	30	95	
v_j					

Table 5b.—XYZ Sawmill Company transportation problem.

	Mill A	Mill B	Mill C	Supply	u_i
Site 1	32 20	60	200	20	
Site 2	40 10	68 20	80	30	
Site 3	120	104 15	60 30	45	
Demand	30	35	30	95	
v_j					

Table 5c.—XYZ Sawmill Company transportation problem.

	Mill A	Mill B	Mill C	Supply	u_i
Site 1	32 20	60	200	20	–8
Site 2	40 10	68 20	80	30	0
Site 3	120	104 15	60 30	45	36
Demand	30	35	30	95	
v_j	40	68	–6		

7.6 Summary

The Transportation models play a significant part in logistics and supply chain management for reducing cost and improving service. The goal is to find the most cost effective way to transport the goods. There are various approaches that can be used to solve a transportation problem. The transportation problem is only a special topic of the linear programming problems. It would be a rare instance when a linear programming problem would actually be solved by hand. There are too many computers around and too many LP software programs to justify spending time for manual solution. Although transportation problem solving methods are developed to solve problems in transporting goods from one location to another, the transportation method can also be used to solve other problems. The only thing required is to formulate the problem in transportation form.

7.7 References

Zulkipli Ghazali, M. Amin Abd Majid and Mohd Shazwani, 2012. Optimal Solution of Transportation Problem Using Linear Programming: A Case of a Malaysian Trading Company. Journal of Applied Sciences, 12:2430-2435.DOI:10.3923/jas.2012.2430.2435URL: http://scialert.net/abstract/?doi=jas.2012.2430.2435

Retrieved February 28, 2013

 Int. J. Contemp. Math. Sciences, Vol. 5, 2010, no. 28, 1385 - 1395 Fourier Method for Solving Transportation International Mathematical   Forum, Vol. 6, 2011, no. 40, 1983 – 1992. http://www.m-hikari.com/ijcms-2010/25-28- 2010/pandianIJCMS25-28-2010.pdf

   Retrieved April 1, 2013

P. Pandian and D. Anuradha, Floating Point Method for Solving Transportation Problems with Additional Constraints. Department of Mathematics, School of Advanced Sciences / VIT University, Vellore-14, Tamilnadu, India. Retrieved March 22, 2013

Computer Science & Engineering: An International Journal (CSEIJ), Vol.2, No.5, October 2012. Retrieved April 2, 2013

http://www.slideshare.net/AnkitBist/transportation-problem-10629516

   Retrieved March 14, 2013

D. Dutta and A. Satyanarayana Murthy, Fuzzy transportation problem with additional restrictions, ARPN Journal of Engineering and Applied Sciences, Vol. 5, No.2, 2010. Retrieved January 30, 2013

http://www.mbaknol.com/management-science/initial-basic-feasible-solution-of-a-transportaion-problem/. Retrieved March 3, 2013 http://www.faculty.kfupm.edu.sa/CEM/.../The-Transportation-Problem.ppt.You +I'd this publicly. Retrieved January 30, 2013

Undo

   http://staff.aub.edu.lb/~bm05/ENMG500/Set_7_TP_a.pdf. Retrieved March 1, 2013

http://home.ubalt.edu/ntsbarsh/opre64a/partviii.html Retrieved February 25, 2013

http://twiki.esc.auckland.ac.nz/twiki/pub/OpsRes/CosmicComputersSolverStudio/cosmic_network.jpg . Retrieved February 15, 2013

 J. Reeb and S. Leavengood, Transportation Problem: A Special Case For Linear Programming Problems, Performance Excellence In The Wood Products Industry, Operations Research, 2002. Retrieved February 20, 2013

S. Chopra and P. Meindl, Supply Chain Management, Strategy,   Planning And Operation (Forth Edition) Retrieved February 28, 2013 You +I'd this publicly.

Lecture 8

Power Point Handouts

Assignment & Transshipment Problems

Dr. MD Sarder

Assignment Problem

- A <u>network model</u> is one which can be represented by a set of nodes, a set of arcs, and functions (e.g. costs, supplies, demands, etc.) associated with the arcs and/or nodes. AP is an example of a network problem.

- This method can solve a special form of LP problem, including the classical <u>assignment problem</u>, with these typical characteristics:
    - is a special case of a transportation problem
    - the right-hand sides of constraints are all 1
    - the signs of the constraints are = rather than $\leq$ or $\geq$
    - the value of all decision variables is either 0 or 1

Assignment Problem

- An <u>assignment problem</u> seeks to minimize the total cost assignment of m workers to m jobs, given that the cost of worker i performing job j is c_{ij}.
- It assumes all workers are assigned and each job is performed.
- An assignment problem is a special case of a transportation problem in which all supplies and all demands are equal to 1; hence assignment problems may be solved as linear programs.
- The network representation of an assignment problem with three workers and three jobs is shown on the next slide.

Assignment Problem

- Network Representation

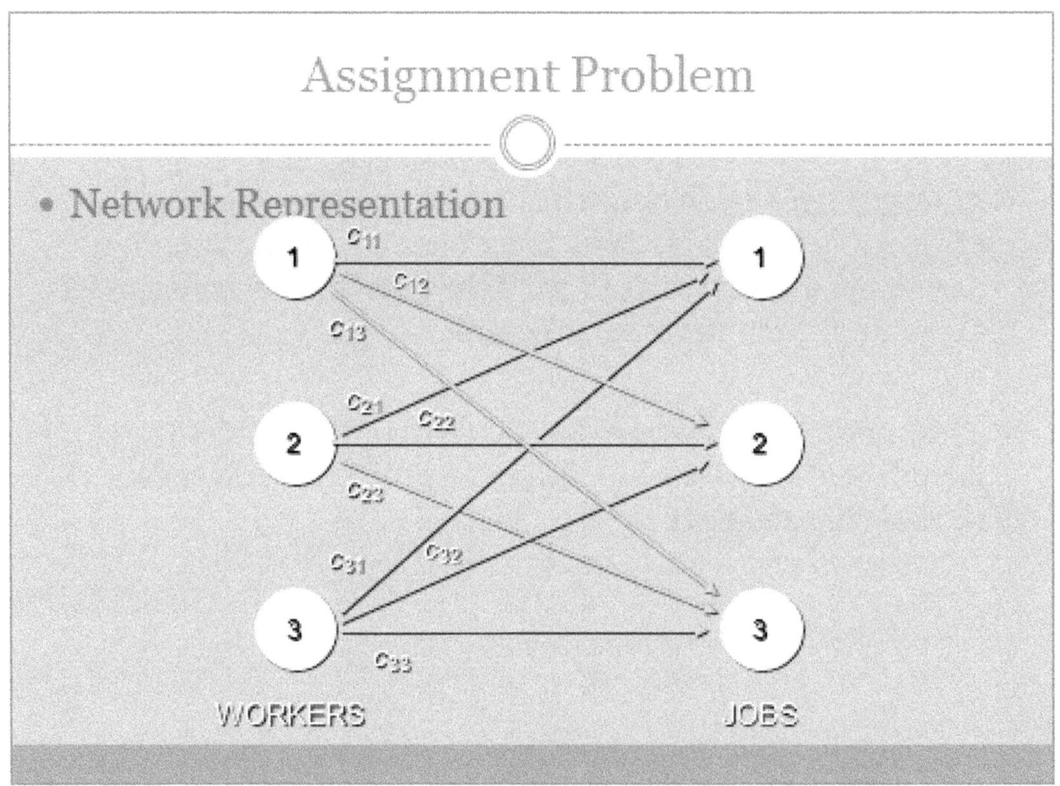

Assignment Problem

- Linear Programming Formulation

$$\text{Min}_{i\,j}\ \Sigma\Sigma c_{ij}x_{ij}$$

$$\text{s.t.}\ \Sigma_j x_{ij} = 1 \qquad \text{for each resource (row) } i$$

$$\Sigma_i x_{ij} = 1 \qquad \text{for each job (column) } j$$

$$x_{ij} = 0 \text{ or } 1 \quad \text{for all } i \text{ and } j.$$

- Note: A modification to the right-hand side of the first constraint set can be made if a worker is permitted to work more than 1 job.

Hungarian Method

- The <u>Hungarian method</u> solves minimization assignment problems with m workers and m jobs.
- Special considerations can include:
  - number of workers does not equal the number of jobs — add dummy workers/jobs with 0 assignment costs as needed
  - worker i cannot do job j — assign $c_{ij} = +M$
  - maximization objective — create an opportunity loss matrix subtracting all profits for each job from the maximum profit for that job before beginning the Hungarian method

Hungarian Method

- Step 1: For each row, subtract the minimum number in that row from all numbers in that row.
- Step 2: For each column, subtract the minimum number in that column from all numbers in that column.
- Step 3: Draw the minimum number of lines to cover all zeroes. If this number = m, STOP — an assignment can be made.
- Step 4: Determine the minimum uncovered number (call it d).
  - Subtract d from uncovered numbers.
  - Add d to numbers covered by two lines.
  - Numbers covered by one line remain the same.
  - Then, GO TO STEP 3.

Hungarian Method

- Finding the Minimum Number of Lines and Determining the Optimal Solution
  - Step 1: Find a row or column with only one unlined zero and circle it. (If all rows/columns have two or more unlined zeroes choose an arbitrary zero.)
  - Step 2: If the circle is in a row with one zero, draw a line through its column. If the circle is in a column with one zero, draw a line through its row. One approach, when all rows and columns have two or more zeroes, is to draw a line through one with the most zeroes, breaking ties arbitrarily.
  - Step 3: Repeat step 2 until all circles are lined. If this minimum number of lines equals m, the circles provide the optimal assignment.

Example 1: AP

A contractor pays his subcontractors a fixed fee plus mileage for work performed. On a given day the contractor is faced with three electrical jobs associated with various projects. Given below are the distances between the subcontractors and the projects.

		Project		
		A	B	C
	Westside	50	36	16
Subcontractors	Federated	28	30	18
	Goliath	35	32	20
	Universal	25	25	14

How should the contractors be assigned to minimize total costs?

Note: There are four subcontractors and three projects. We create a dummy project Dum, which will be assigned to one subcontractor (i.e. that subcontractor will remain idle)

Example 1: AP

- Network Representation (note the dummy project)

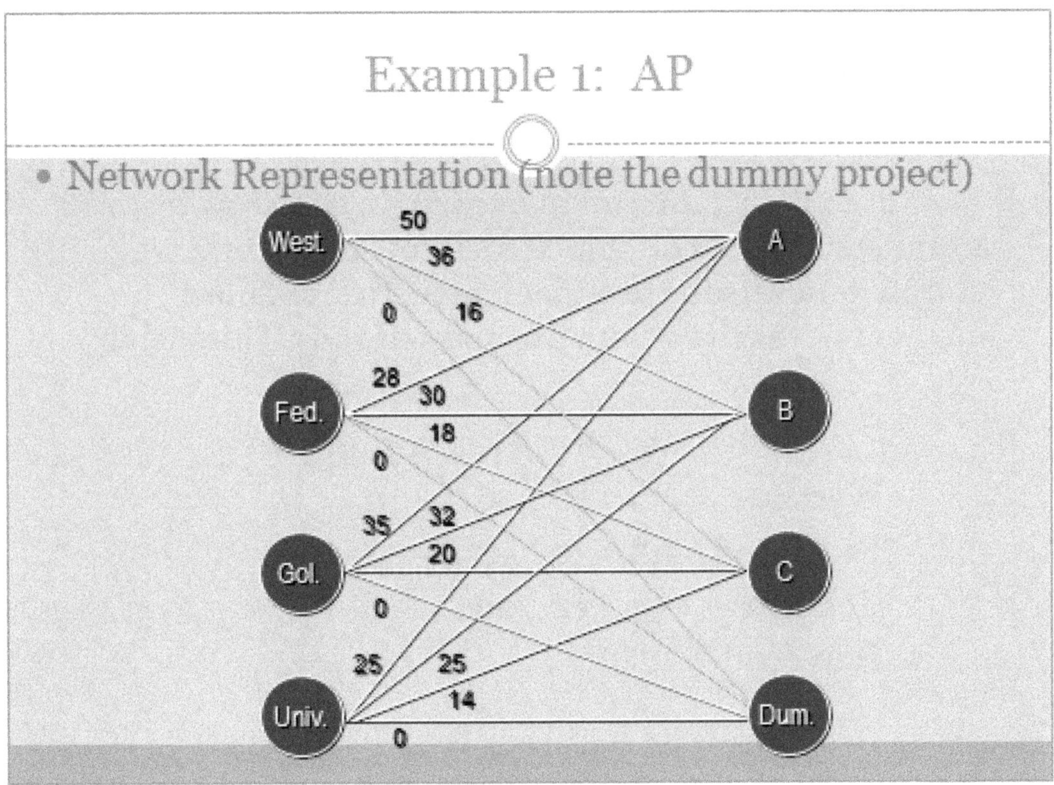

- Initial Tableau Setup

Since the Hungarian algorithm requires that there be the same number of rows as columns, add a Dummy column so that the first tableau is (the smallest elements in each row are marked red):

	A	B	C	Dummy
Westside	50	36	16	0
Federated	28	30	18	0
Goliath	35	32	20	0
Universal	25	25	14	0

Example 1: AP

- Step 1: Subtract minimum number in each row from all numbers in that row. Since each row has a zero, we simply generate the original matrix (the smallest elements in each column are marked red). This yields:

	A	B	C	Dummy
Westside	50	36	16	0
Federated	28	30	18	0
Goliath	35	32	20	0
Universal	25	25	14	0

- Step 2: Subtract the minimum number in each column from all numbers in the column. For A it is 25, for B it is 25, for C it is 14, for Dummy it is 0. This yields:

	A	B	C	Dummy
Westside	25	11	2	0
Federated	3	5	4	0
Goliath	10	7	6	0
Universal	0	0	0	0

Example 1: AP

- Step 3: Draw the minimum number of lines to cover all zeroes (called minimum cover). Although one can "eyeball" this minimum, use the following algorithm. If a "remaining" row has only one zero, draw a line through the column. If a remaining column has only one zero in it, draw a line through the row. Since the number of lines that cover all zeros is 2 < 4 (# of rows), the current solution is not optimal.

	A	B	C	Dummy
Westside	25	11	(2)	0
Federated	3	5	4	0
Goliath	10	7	6	0
Universal	0	0	0	0

- Step 4: The minimum uncovered number is 2 (circled).

- Step 5: Subtract 2 from uncovered numbers; add 2 to all numbers at line intersections; leave all other numbers intact. This gives:

	A	B	C	Dummy
Westside	23	9	0	0
Federated	1	3	2	0
Goliath	8	5	4	0
Universal	0	0	0	2

- Step 3: Draw the minimum number of lines to cover all zeroes. Since 3 (# of lines) < 4 (# of rows), the current solution is not optimal.

	A	B	C	Dummy
Westside	23	9	0	0
Federated	1	3	2	0
Goliath	8	5	4	0
Universal	0	0	0	2

- Step 4: The minimum uncovered number is 1 (circled).

- Step 5: Subtract 1 from uncovered numbers. Add 1 to numbers at intersections. Leave other numbers intact. This gives:

	A	B	C	Dummy
Westside	23	9	0	1
Federated	0	2	1	0
Goliath	7	4	3	0
Universal	0	0	0	3

Find the minimum cover:

	A	B	C	Dummy
Westside	23	9	0	1
Federated	0	2	1	0
Goliath	7	4	3	0
Universal	0	0	0	3

- Step 4: The minimum number of lines to cover all 0's is four. Thus, the current solution is optimal (minimum cost) assignment.

The optimal assignment occurs at locations of zeros such that there is exactly one zero in each row and each column:

	A	B	C	Dummy
Westside	23	9	0	1
Federated	0	2	1	0
Goliath	7	4	3	0
Universal	0	0	0	3

The optimal assignment is (go back to the original table for the distances):

Subcontractor	Project	Distance
Westside	C	16
Federated	A	28
Universal	B	25
Goliath	(unassigned)	

Total Distance = 69 miles

Example 1: AP via LP

- In our example the LP formulation is:

$$\text{Min } z = 50x_{11} + 36x_{12} + 16x_{13} + 0x_{14} + 28x_{21} + 30x_{22} + 18x_{23} + 0x_{24} +$$
$$35x_{31} + 32x_{32} + 20x_{33} + 0x_{34} + + 25x_{41} + 25x_{42} + 14x_{43} + 0x_{44}$$

s.t.

$$x_{11} + x_{12} + x_{13} + x_{14} = 1 \ (\text{row } 1)$$
$$x_{21} + x_{22} + x_{23} + x_{24} = 1 (\text{row } 2)$$
$$x_{31} + x_{32} + x_{33} + x_{34} = 1 (\text{row } 3)$$
$$x_{41} + x_{42} + x_{43} + x_{44} = 1 (\text{row } 4)$$
$$x_{11} + x_{21} + x_{31} + x_{41} = 1 \ (\text{column } 1)$$
$$x_{12} + x_{22} + x_{32} + x_{42} = 1 (\text{column } 2)$$
$$x_{13} + x_{23} + x_{33} + x_{43} = 1 \ (\text{column } 3)$$
$$x_{14} + x_{24} + x_{34} + x_{44} = 1 \ (\text{column } 4)$$
$$x_{ij} >= 0 \text{ for } i = 1, 2, 3, 4 \text{ and } j = 1, 2, 3, 4 \ (\text{nonnegativity})$$

Example 1: AP via LP

- The solver formulation is:

The Assignment Problem

Distances From Contractors to Projects at:

Contractor	A	B	C	Dummy
West Side	50	36	18	0
Federated	28	30	18	0
Goliath	35	32	20	0
Universal	25	25	14	0

Assignment of Contractors to Projects at:

Contractor	A	B	C	Dummy	Assigned	Available
West Side	0	0	0	0	0	1
Federated	0	0	0	0	0	1
Goliath	0	0	0	0	0	1
Universal	0	0	0	0	0	1
Assigned	0	0	0	0		
Capacity	1	1	1	1		

Total Distance: 0

Example 1: AP via LP

- The solver solution is:

The Assignment Problem

Distances From
Contractors to Projects at:

Contractor	A	B	C	Dummy
West Side	50	38	18	0
Federated	28	30	18	0
Goliath	35	32	20	0
Universal	25	25	14	0

Assignment of Contractors
to Projects at:

Contractor	A	B	C	Dummy	Assigned	Available
West Side	0	0	1	0	1	1
Federated	1	0	0	0	1	1
Goliath	0	0	0	1	1	1
Universal	0	1	0	0	1	1
Assigned	1	1	1	1		
Capacity	1	1	1	1		

Total Distance: 88

Transshipment Problem

- <u>Transshipment problems</u> are transportation problems in which a shipment may move through intermediate nodes (transshipment nodes) before reaching a particular destination node.
- Transshipment problems can be converted to larger transportation problems and solved by a special transportation program.
- Transshipment problems can also be solved by general purpose linear programming codes.
- The network representation for a transshipment problem with two sources, three intermediate nodes, and two destinations is shown on the next slide.

Transsshipment Problem

- Network Representation

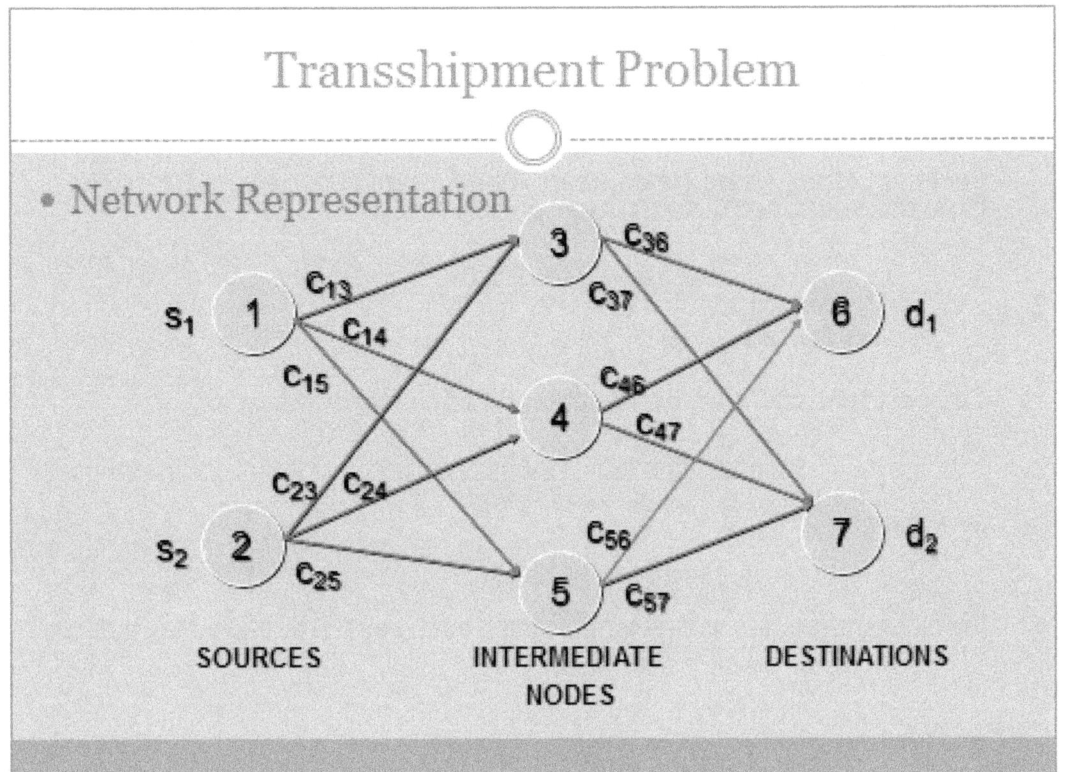

Example 1: Thomas & Washburn

Thomas Industries and Washburn Corporation supply three firms (Zrox, Hewes, Rockwright) with customized shelving for its offices. They both order shelving from the same two manufacturers, Arnold Manufacturers and Supershelf, Inc.

Currently weekly demands by the users are 50 for Zrox, 60 for Hewes, and 40 for Rockwright. Both Arnold and Supershelf can supply at most 75 units to its customers.

Additional data is shown on the next slide.

Example 1: Thomas & Washburn

Because of long standing contracts based on past orders, unit costs from the manufacturers to the suppliers are:

	Thomas	Washburn
Arnold	5	8
Supershelf	7	4

The cost to install the shelving at the various locations are:

	Zrox	Hewes	Rockwright
Thomas	1	5	8
Washburn	3	4	4

Find the quantities to be shipped from each source to each destination to minimize the total shipping cost.

Example 1: Thomas & Washburn

- Network Representation

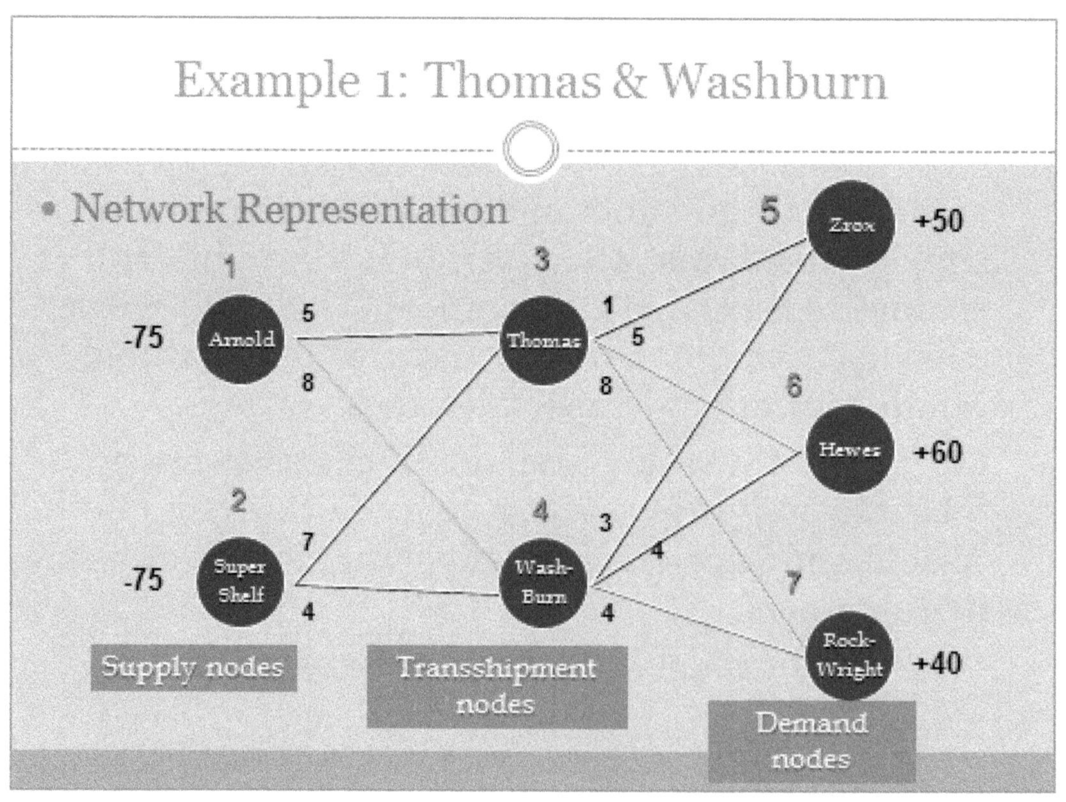

Example 1: Thomas & Washburn

- Linear Programming Formulation
  - Decision Variables Defined

x_{ij} = amount shipped from manufacturer i to supplier j

x_{jk} = amount shipped from supplier j to customer k

$$\text{where} \quad i = 1 \,(\text{Arnold}), 2 \,(\text{Supershelf})$$
$$j = 3 \,(\text{Thomas}), 4 \,(\text{Washburn})$$
$$k = 5 \,(\text{Zrox}), 6 \,(\text{Hewes}), 7 \,(\text{Rockwright})$$

  - Objective Function Defined

Minimize Overall Shipping Costs:

$$\text{Min} \quad 5x_{13} + 8x_{14} + 7x_{23} + 4x_{24} + 1x_{35} + 5x_{36} + 8x_{37}$$
$$+ \, 3x_{45} + 4x_{46} + 4x_{47}$$

Example 1: Thomas & Washburn

- Constraints Defined

Amount Out of Arnold:	$x_{13} + x_{14} \leq 75$
Amount Out of Supershelf:	$x_{23} + x_{24} \leq 75$
Amount Through Thomas:	$x_{13} + x_{23} - x_{35} - x_{36} - x_{37} = 0$
Amount Through Washburn:	$x_{14} + x_{24} - x_{45} - x_{46} - x_{47} = 0$
Amount Into Zrox:	$x_{35} + x_{45} \geq 50$
Amount Into Hewes:	$x_{36} + x_{46} \geq 60$
Amount Into Rockwright:	$x_{37} + x_{47} \geq 40$

Non-negativity of Variables: $x_{ij} \geq 0$, for all i and j.

Example 1: Thomas & Washburn problem via LP

- ## The solver formulation is:

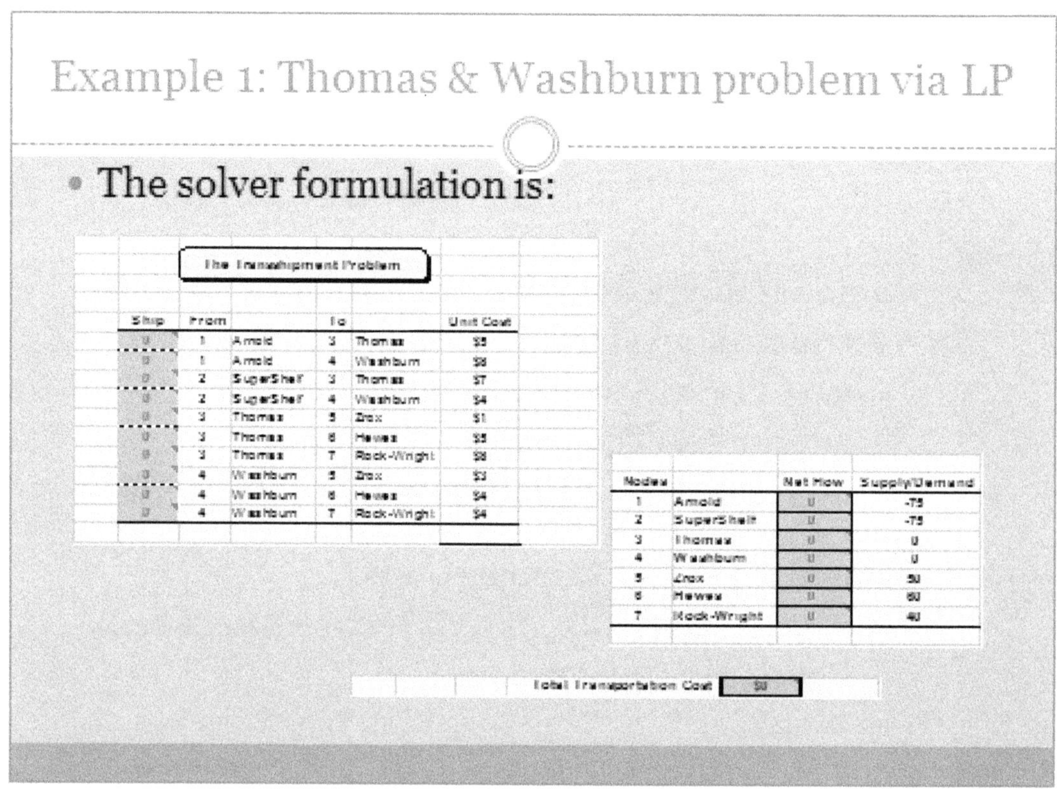

Example 1: Thomas & Washburn problem via LP

- ## The solver solution is:

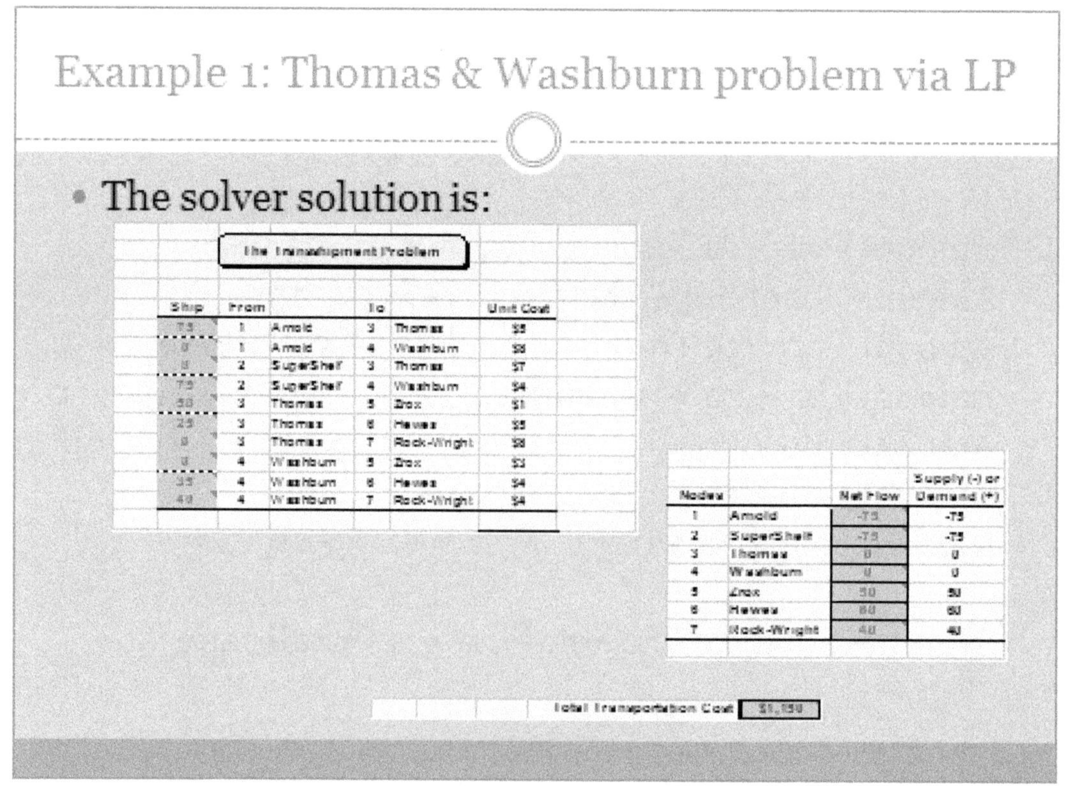

Example 1: Thomas & Washburn problem via NP

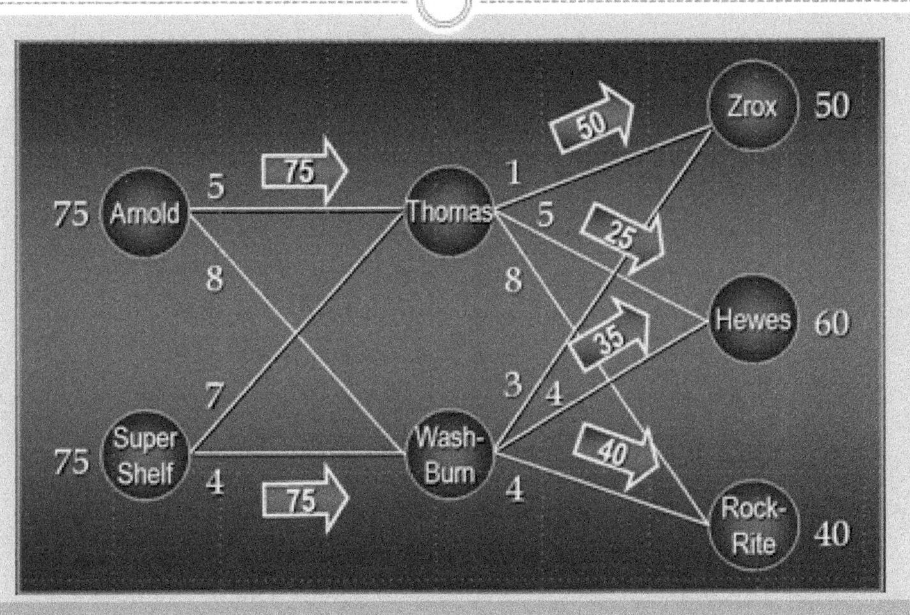

Example 2: The Bavarian Motors Company (BMC)

BMC manufactures (very) expensive sports luxury cars in Germany and ships them for sale in the US to Newark and Jacksonville. From these entry ports the cars can be transported to distributors located in Boston, Columbus, Atlanta, Richmond and Mobile. There are 200 cars available for shipment in Newark and 300 in Jacksonville. The distributors ordered the following quantities of cars:

- Boston: 100
- Columbus: 60
- Atlanta: 170
- Richmond: 80
- Mobile: 70.

The shipping costs per unit for are shown in the next slide. You are retained (for a large consulting fee) to determine shipping quantities to minimize the total shipping cost.

From		To		Unit Cost
1	Newark	2	Boston	$30
1	Newark	4	Richmond	$40
2	Boston	3	Columbus	$50
3	Columbus	5	Atlanta	$35
5	Atlanta	3	Columbus	$40
5	Atlanta	4	Richmond	$30
5	Atlanta	6	Mobile	$35
6	Mobile	5	Atlanta	$25
7	Jacksonville	4	Richmond	$50
7	Jacksonville	5	Atlanta	$45
7	Jacksonville	6	Mobile	$50

- Network Representation

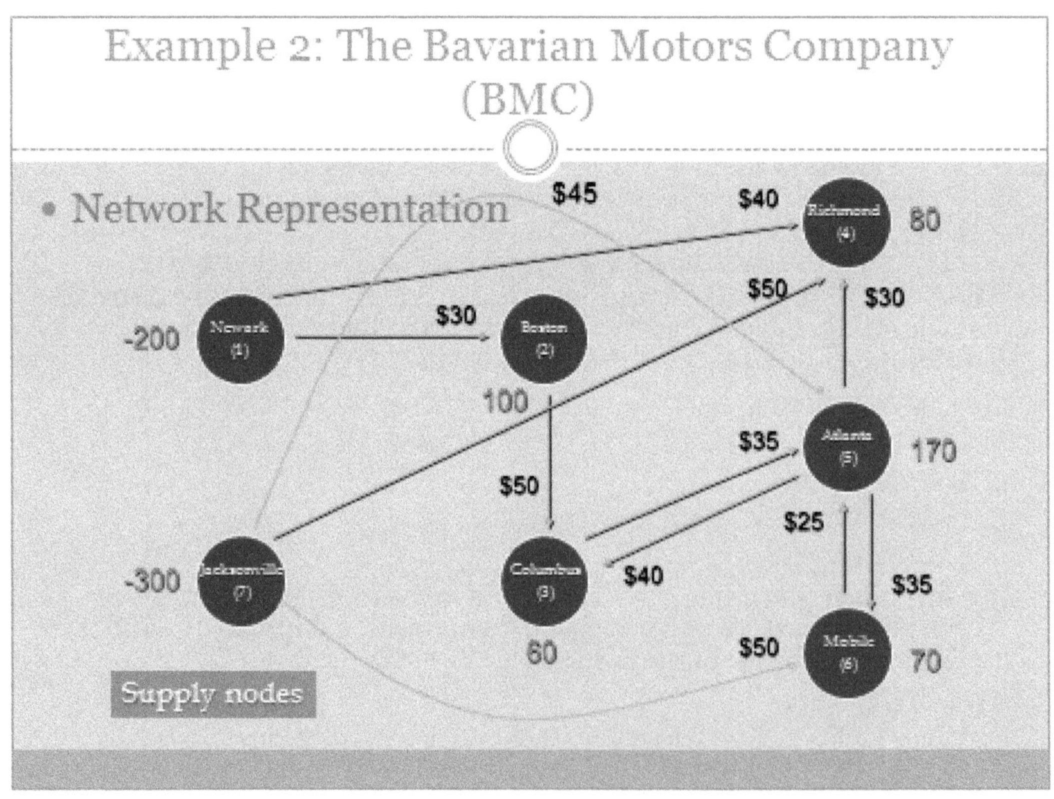

Example 2: The Bavarian Motors Company (BMC)

- Linear Programming Formulation:

  - Decision Variables Defined

    x_{ij} = amount shipped from source i to destination j
    (i, j = 1 through 7)

  - Objective Function Defined

    Minimize Overall Shipping Costs:

$$\text{Min} \quad 30x_{12} + 40x_{14} + 50x_{23} + 35x_{35} + 40x_{53} + 30x_{54} +$$
$$35x_{56} + 25x_{65} + 50x_{74} + 45x_{75} + 50x_{76}$$

Example 2: The Bavarian Motors Company (BMC)

- Constraints Defined

  Flow constraint for node 1: $\quad -x_{12} - x_{14} \geq -200$
  Flow constraint for node 2: $\quad x_{12} - x_{23} \geq 100$
  Flow constraint for node 3: $\quad x_{23} + x_{53} - x_{35} \geq 60$
  Flow constraint for node 4: $\quad x_{14} + x_{54} + x_{74} \geq 80$
  Flow constraint for node 5:

  $$x_{35} + x_{65} + x_{75} - x_{53} - x_{54} - x_{56} \geq 170$$

  Flow constraint for node 6: $\quad x_{56} + x_{76} - x_{65} \geq 70$
  Flow constraint for node 7: $\quad -x_{74} - x_{75} - x_{76} \geq -300$

  Non-negativity of variables: $x_{ij} \geq 0$, for all i and j.

- ## The solver formulation is:

Ship	From		To		Unit Cost		Nodes		Net Flow	Supply/Demand
0	1	Newark	2	Boston	$30		1	Newark	0	-200
0	1	Newark	4	Richmond	$40		2	Boston	0	100
0	2	Boston	3	Columbus	$50		3	Columbus	0	60
0	3	Columbus	5	Atlanta	$35		4	Richmond	0	80
0	5	Atlanta	3	Columbus	$40		5	Atlanta	0	170
0	5	Atlanta	4	Richmond	$30		6	Mobile	0	70
0	5	Atlanta	6	Mobile	$35		7	Jacksonville	0	-300
0	6	Mobile	5	Atlanta	$25					
0	7	Jacksonville	4	Richmond	$50					
0	7	Jacksonville	5	Atlanta	$45					
0	7	Jacksonville	6	Mobile	$50					

Total Transportation Cost: $0

Example 2: The Bavarian Motors Company (BMC)

- ## The solver solution is:

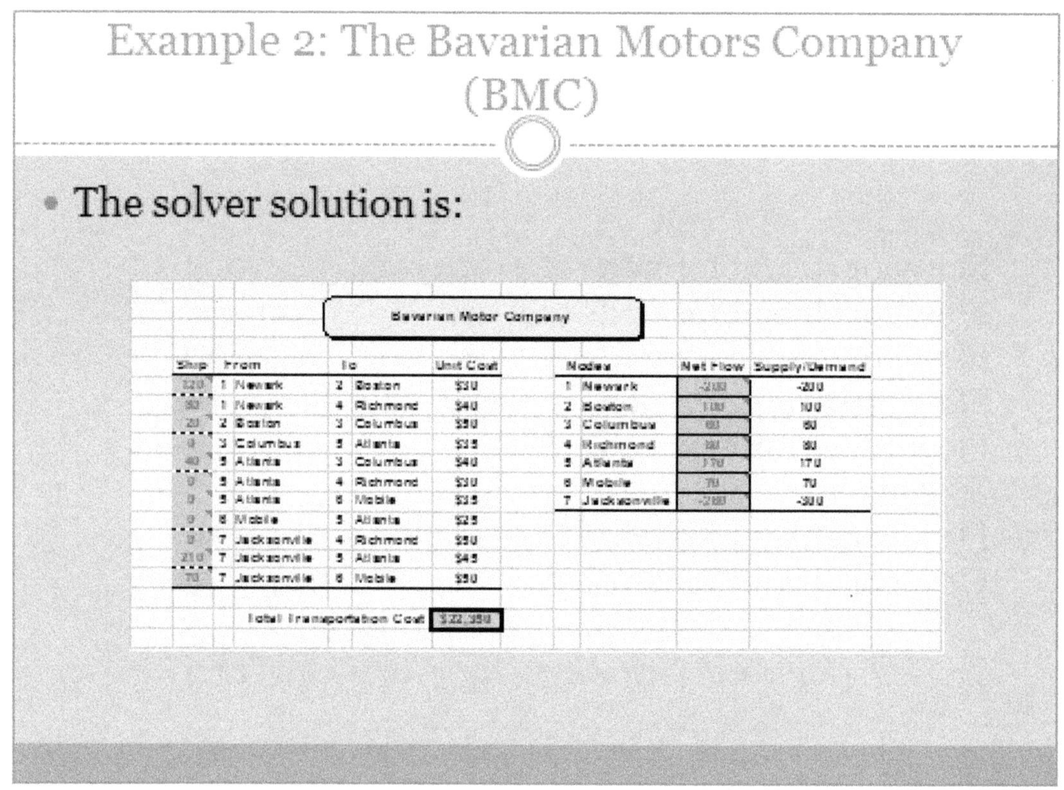

Ship	From		To		Unit Cost		Nodes		Net Flow	Supply/Demand
120	1	Newark	2	Boston	$30		1	Newark	-200	-200
50	1	Newark	4	Richmond	$40		2	Boston	100	100
20	2	Boston	3	Columbus	$50		3	Columbus	60	60
0	3	Columbus	5	Atlanta	$35		4	Richmond	80	80
40	5	Atlanta	3	Columbus	$40		5	Atlanta	170	170
0	5	Atlanta	4	Richmond	$30		6	Mobile	70	70
0	5	Atlanta	6	Mobile	$35		7	Jacksonville	-200	-300
0	6	Mobile	5	Atlanta	$25					
0	7	Jacksonville	4	Richmond	$50					
210	7	Jacksonville	5	Atlanta	$45					
70	7	Jacksonville	6	Mobile	$50					

Total Transportation Cost: $22,350

LECTURE 8

READING MATERIALS

8.0 Introduction

In today's business environment, managers are often called upon to make very complicated decisions with millions of dollars riding on the outcome. To help them achieve the most accurate decision, managers use linear programming. Linear programming is a management science technique used in business today and is instrumental in solving a wide range of management problems by maximizing or minimizing quantity, usually a cost or profit, subject to constraints (Linear Programming, 2010).

Linear programming was initially developed, along with many other management science techniques, as a discipline in the 1940's during and shortly after World War II. Its founder, George Danzig, developed the procedures for solving linear programming problems while working on ways to minimize the total costs of allocating supplies for Air Force troops during the war (Overton, 1997). Its development quickly accelerated after World War II as many industries found the technique valuable and began adopting it as a standard tool to allocate resources in an optimal way.

Today, linear programming is used to solve decision problems in many interesting and important applications which may contain thousands of variables. Some of the applications for its use include:

- product mix planning
- distribution networks
- truck routing
- staff scheduling
- financial portfolios
- corporate restructuring

In business and industry today, linear programming is so effective it is said to account for as much as 90 percent of all computing time for business management decisions (Linear Programming, 2010). In a relatively short period of time, it has changed the way business managers make decisions, from that of guesswork and intuition, to the use of an algorithm based on available data that accurately produces optimal decisions.

8.1 Literature Review

Linear programming is a mathematical method which aims to achieve the objective of meeting a desired goal of highest profit or lowest cost with efficient allocation of limited resources to known activities. Linear programming has been successfully applied to various fields of study. It not only can be widely used in business and economic areas, but also can be utilized in engineering situation. Some industries using Linear Programming include transportation, telecommunications, production, energy, and manufacturing companies. To the extent, linear programming has proved useful in modeling diverse types of problems in planning, routing, scheduling assignment and design.

David (1982) stated operational research is a mathematical method developed to solve problems related to tactical and strategic operations. Its origins show its application in the decision-making process of business analysis, mainly regarding the best use for short funds. This shortage of funds is a characteristic of hyper-competitive environments. Although the practical application of a mathematical model is wide and complex, it will provide a set of results that enable the elimination of a part of the subjectivism that exists in the decision-making process as to the choice of action alternatives (Bierman and Bonini, 1973).

Linear Programming is a considerable field of optimization for several reasons. Many practical problems in operations problems in operations research can be expressed as linear programming problems. After that, the terminologies, such as transportation/assignment problems, and allocation problems, have become a standard in these contexts.

In an assignment problem, one is looking to find an assignment, or matching, between the elements of two (or more) sets, such that the total cost of all matched pairs is minimized. Depending on the particular structure of the sets being matched, the form of the cost function, the matching rule, and so on, the assignment problems are categorized into linear, quadratic, bottleneck, multidimensional, etc. The assignment problems can be stated in a variety of forms, including mathematical programming, combinatorial, or graph-theoretic formulations, and constitute one of the most important and fundamental objects in computer science, operations research, and discrete mathematics. Besides these areas of their "natural habitat," assignment problems have found numerous applications in other disciplines of science and engineering, including chemistry, biology, physics, archeology, electrical engineering, sports, and others (for a comprehensive review of the subject of assignment problems, their formulations and applications, see, for instance, Pardalos and Pitsoulis, 2000; Burkard, 2002; Pentico, 2007, and references therein).

For a standard assignment problem, only the cost or profit from each possible assignment is considered in the problem formulation. However, in real applications, for each possible assignment several kinds of input resources are usually needed in an assignment problem. Moreover, decision-makers can have several different objectives to achieve for each possible assignment, and the ways to achieve these objectives may conflict with each other. For example, for solving the assignment problem of allocating cross-trained workers in a multidepartment and labor-intensive environment, Campbell and Diaby (2002) pointed out that demand levels in the various departments and the capabilities of the available workers should be considered as the inputs, and that the assignment outcomes can affect quality of service and employee satisfaction. They emphasized that effective cross-utilization of manpower is of practical importance in certain professions, such as nursing.

8.2 Mathematical formula of solving assignment program using LP

As we mentioned before, linear programming and assignment problems have been successfully apply in different areas. In terms of the linear assignment problem, it is one of the basic and fundamental models in operations research, computer science and discrete mathematics. In its most familiar interpretation, it solves the problem of finding an assignment of n workers to n jobs which has the lowest overall cost, using C_{ij} to represent the cost of assigning worker i to task j. Except the straightforward applications, this linear assignment problem often arises as a part of some optimization problems, such as multidimensional assignment problem, quadratic assignment problem, traveling salesman problem, supply chain network design, etc.

The mathematical formulation has the form as below:

$$L_n = \min \sum_{i=1}^{n} \sum_{j=1}^{n} C_{ij} X_{ij}$$

s.t.

$$\sum_{i=1}^{n} X_{ij} = 1, \quad j = 1, \dots, n,$$

$$\sum_{j=1}^{n} X_{ij} = 1, \quad i = 1, \dots, n,$$

According to the opinion of Pavlo and Panos (2007), The decision variables X_{ij} can be taken as either binary: $X_{ij} \in \{0,1\}$, or non-negative: $X_{ij} \geq 0$, leading to a an integer programming or linear programming formulations respectively. In the graph-theoretical setting, this problem corresponds to finding a minimum cost perfect matching in an edge-weighed bipartite graph; another useful interpretation of rows and columns of the cost matrix $C = (C_{ij})$ that minimizes the sum of the elements on the diagonal. The latter observation leads to the permutation formulation of the linear assignment problem.

8.3 Case Study

8.3.1 Problem Description

A manufacturer company is considering transporting products from different regions to different markets. Instead of setting up facility in each region, they want to consolidate plants in a few regions. Although it improves economics of scale, it will increases transportation cost and duties. There variable production, inventory, and transportation cost (including tariffs and duties) of producing in one region to meet demand in each individual market demand. Fixed cost is also associated with facilities, transportation, and inventories at each facility. The company is considering two different plant sizes in each location, Low-capacity plants and high-capacity plants respectively. Our aim is to find the lowest cost network.

Inputs are shown as follow:

n = no. of potential plant locations for each level of capacity

m = no. of markets or demand points

D_j = annual demand from market j

K_j = potential capacity of plant i

f_i = annualized fixed cost of keeping factory i open

C_{ij} = cost of producing and shipping one unit from factory i to market j

We assume that all demand must be met and taxes on earnings are ignored, so the model focus on the minimizing the cost of meeting global demand. Define the following decision variables:

y_i = I if plant i is open, 0 otherwise.

x_{ij} = quantity shipped from plant i to market j

The fixed cost will be:

$$\sum_{i=1}^{n} f_i y_i$$

The variable cost will be:

$$\sum_{i=1}^{n} \sum_{j=1}^{m} c_{ij} x_{ij}$$

The problem is formulated as:

$$\text{Min} \sum_{i=1}^{n} f_i y_i + \sum_{i=1}^{n} \sum_{j=1}^{m} c_{ij} x_{ij}$$

Subject to

$$\sum_{i=1}^{n} x_{ij} = D_j \qquad \text{for } j = 1, \ldots, m \qquad (1)$$

$$\sum_{j=1}^{m} x_{ij} \leq K_i y_i \qquad \text{for } i = 1, \ldots, n \qquad (2)$$

$$y_i \in \{0,1\} \qquad \text{for } i = 1, \ldots, n, x_{ij} \geq 0 \qquad (3)$$

8.3.2 Solution Method

The first step is to collect the data in a form that could be used for a quantitative model. In this company, annual demand of each market is shown in the table, as well as the variable production, inventory and transportation cost (including tariffs and duties) of producing in one region to meet demand in each individual region. All costs are in thousands of dollars. Here is the specific data:

Demand Region (Production and Transportation Cost per 1,000,000 Units)

Supply Region	Market 1	Market 2	Market 3	Market 4	Market 5
Demand	**11**	**9**	**16**	**17**	**9**
Factory 1	75	92	110	129	119
Factory 2	119	73	107	101	107
Factory 3	103	107	85	121	115
Factory 4	117	128	93	61	74
Factory 5	145	102	105	109	69

The company is considering two different plant sizes in each location, low-capacity plants which can produce 10 million units one year, while high-capacity plants which can produce 20 million units. All fixed costs are annualized. The fixed costs are shown in the following table:

Supply Region	Fixed Cost	Low Capacity	Fixed Cost	High Capacity
Factory 1	5,000	10	7,500	20
Factory 2	4,700	10	7,050	20
Factory 3	6,300	10	9,450	20
Factory 4	4,200	10	6,300	20
Factory 5	4,600	10	6,900	20

The second step is to use Solver tool in Excel. According to the given data, we can identify cells corresponding to each decision variable as shown in the following table. The decision variables x_{ij} determine the amount produced in a factory and shipped to a market region, whereas decision variables y_i determine plants of different levels of capacity. Initially, all decision variables are set to be 0.

Supply Region	Market 1	Market 2	Market 3	Market 4	Market 5	Low Capacity	High Capacity
Factory 1	0	0	0	0	0	0	0
Factory 2	0	0	0	0	0	0	0
Factory 3	0	0	0	0	0	0	0
Factory 4	0	0	0	0	0	0	0
Factory 5	0	0	0	0	0	0	0

The next step is to construct cells for the constraints in Equations (1) and (2). The constraint in (1) requires the demand at each regional market be satisfied. The constraint (2) states that on plant can supply more than its capacity, clearly, the capacity is 0 if the plant is closed and K_i if it is open. The constraint (3) enforces that each plant is either open ($y_i = 1$) or closed ($y_i = 0$). The solution identifies the plants that are to be kept open, their capacity, and the allocation of regional demand to these plants.

$$\sum_{i=1}^{n} x_{ij} = D_j \qquad \text{for } j = 1, \dots, m \qquad (1)$$

$$\sum_{j=1}^{m} x_{ij} \leq K_i y_i \qquad \text{for } i = 1, \dots, \qquad (2)$$

$$y_i \in \{0,1\} \qquad \text{for } i = 1, \dots, n, x_{ij} \geq 0 \qquad (3)$$

The constraint cells and objective function are shown in the follow table.

Cell	Cell Formula	Equation	Copied to
B41	B5-SUM(B26:B30)	(1)	B41:F41
B34	G26*C17+H26*E17-SUM(B26:F26)	(2)	B34:B38
B46	SUMPRODUCT(B26:F30,B6:F10)+SUMPRODUCT(G26:G30, B17:B21)+SUMPRODUCT(H26:H30,D17:D21)	Objective Function	-

The next step is to use Data Solver to invoke Solver. The objective function measures the total fixed cost plus the variable cost of operating the network.

Here is the final result solved by Excel.

	A	B	C	D	E	F	G	H
23								
24	**Decision Variables**						Low	High
25	*Supply Region*	Market 1	Market 2	Market 3	Market 4	Market 5	Capacity	Capacity
26	Factory 1	0	0	0	0	0	0	0
27	Factory 2	11	9	0	0	0	0	1
28	Factory 3	0	0	0	0	0	0	0
29	Factory 4	0	0	5	17	0	1	1
30	Factory 5	0	0	11	0	9	0	1
31								
32	**Constraints**							
33	Supply region	Capacity						
34	Factory 1	0						
35	Factory 2	0						
36	Factory 3	0						
37	Factory 4	8						
38	Factory 5	0						
39								
40	Demand	Market 1	Market 2	Market 3	Market 4	Market 5		
41	Unmet	0	0	0	0	0		
42								
43								
44	**Objective function**							
45								
46	Cost =	29,694						
47								

8.4 Results

Based on the calculation above, the supply chain team concludes that the lowest- cost network will have facilities located in Factory 4, while the high-capacity demand should be planned in Factory 2, 4 and 5. The plant in Factory 2 meets the demand of Market 1 and 2, whereas Market 3, 4 and 5 demands are met from plants in Factory 3 and 4. The costs associated with a variety of options incorporating different combinations of strategic concerns such as local presence should be evaluated. A suitable regional configuration is then selected, so the assignment problem has been solved using linear programming.

Conclusion

In this chapter we discussed Linear Programming and how it is used to solve assignment problems. We touched upon the origins of linear programming and what industries depend upon LP to solve very complex problems. We demonstrated through a case study how linear programming can be applied to a problem to derive an optimal solution of lowest cost through step by step use of Microsoft Excel Solver software.

However, it is important for managers and those using linear programming to solve problems to understand that linear programming does have its advantages as well as limitations. It is very good at making the best possible use of available resources such as labor and time. It also works well within the production process by highlighting machine efficiency and which machines have significant idle time or create bottlenecks (Lingham, 2012).

The obvious limitations of linear programming are that it applies only to problems where the constraints and objective function is linear, or can be represented in a straight line. In real life situations, this is not always the case. Other factors such as uncertainty, weather conditions, etc. are also not taken into consideration which can cause significant errors. For this reason, managers must continuously update linear programming problems with the latest information available to them

References

Bierman Jr., Bonini H., & Charles P. (1973): Quantitative Analysis for Business Decisions. 4th edition, Richard D. Irwin, Illinois

Burkard, R.E., 2002. Selected topics on assignment problems: Discrete Applied Mathematics 123 (1–3), 257–302.

David Charles. H. (1982): Introduction Concept and Application of Operation Research G.M. Campbell, M. Diaby, Development and evaluation of an assignment heuristic for allocating cross-trained workers, Eur. J. Oper. Res. 138 (2002) 9–12.

Linear Programming. (2010) Available: http://www.netmba.com/operations/lp/.

Lingham, L. (2012). Human Resources/*Please answer these questions*. Retrieved from http://en.allexperts.com/q/Human-Resources-2866/2012/5/please-answer-questions.htm.

Overton, M.. (1997). *Linear Programming*. Available: http://cs.nyu.edu/faculty/overton/g22_lp/encyc/article_web.html.

Pardalos, P.M., Pitsoulis, L. (Eds.), 2000. Nonlinear Assignment Problems: Algorithms and Applications. Kluwer Academic Publishers

Pavlo A. Krokhmal, Panos M. Pardalos, 2007. Random assignment problems. European Journal of Operational Research 194 (2009) 1–17

Pentico, D.W., 2007. Assignment problems: A golden anniversary survey. European Journal of Operational Research 176 (2), 774–793.

Sunil C., Peter M., Dharam V. K., 2013. Supply Chain Management: Strategy, Planning and Operations, 5th Edition, 129-133.

Solving Transshipment Problems using Linear Programming

Glossary of Terms

Node: a connection point, either a redistribution point or a <u>communication endpoint</u> [5]

Transport hub: a place where <u>passengers</u> and <u>cargo</u> are exchanged between vehicles or between <u>transport modes</u> [7]

Transshipment: the <u>shipment</u> of <u>goods</u> or <u>containers</u> to an intermediate destination, and then from there to yet another destination [9]

Constraint: a condition that a solution to an <u>optimization</u> problem is required by the problem itself to satisfy [1]

Optimization Equation: an equation that signifies the selection of a best element (with regard to some criteria) from some set of available alternatives. [4]

Supply: the amount of some product producers are willing and able to sell at a given price all other factors being held constant [6]

Demand: Demand refers to how much (quantity) of a product or service is desired by buyers [2]

Linear Programming: is a mathematical method for determining a way to achieve the best outcome (such as maximum profit or lowest cost) in a given <u>mathematical model</u> for some list of requirements represented as linear relationships [3]

8.1.A The Nature of the Transshipment Problem

Minimizing shipping cost has become an extremely important factor in the operations of businesses today. Companies are consistently looking for ways to reduce costs in an effort to maximize profit. The transshipment problem is a special case of the transportation problem in that when leaving the supply nodes, shipments may move through intermediate nodes or transport hubs, before reaching their final destination. Transshipment is defined as the shipment of goods or containers to an intermediate destination, and then from there to yet another destination. [9]

Figure I shows a pictorial image of the supply nodes, demand, nodes, and intermediate nodes associated with a typical transshipment problem. For each supply node, there will be an associated supply amount, which the plant location or distribution is capable of producing and shipping. Likewise, there will be an associated demand amount for which is needed for the demand location. The intermediate nodes will receive the determined amounts of product from the supply locations, but the products will then ship from the intermediate node locations and to the final demand locations.

The arrows drawn between the nodes represent costs. For each supply node location, there will be an associated cost to ship per product to the intermediate location. The costs will differ from each supply to each intermediate node. Likewise, there is an associated cost for shipping per product from each intermediate location to each demand point. Again, the costs will differ from each intermediate node to each demand location.

The overall objective of the transshipment problem is to examine the supply, intermediate, and demand routes, along with their associated costs and determine the most cost efficient way to ship the products and the optimal amount of products to come from each supply location, through which intermediate location, and finally to the demand location. The following sections in this chapter will describe the formulation and solving of the transshipment problem using linear programming. This chapter will show some examples of how the transshipment problem is used in real life situations and some case study to clarify its usage. It will also cover the basics of linear programming and software available to better enhance the capabilities of solving these types of problems.

8.1.B Examples of Transshipment Problems

Widgetco manufactures widgets at two factories, one in Memphis and one in Denver. The Memphis factory can produce as 150 widgets, and the Denver factory can produce as many as 200 widgets per day. Widgets are shipped by air to customers in LA and Boston. The customers in each city require 130 widgets per day. Because of the deregulation of airfares, Widgetco believes that it may be cheaper first fly

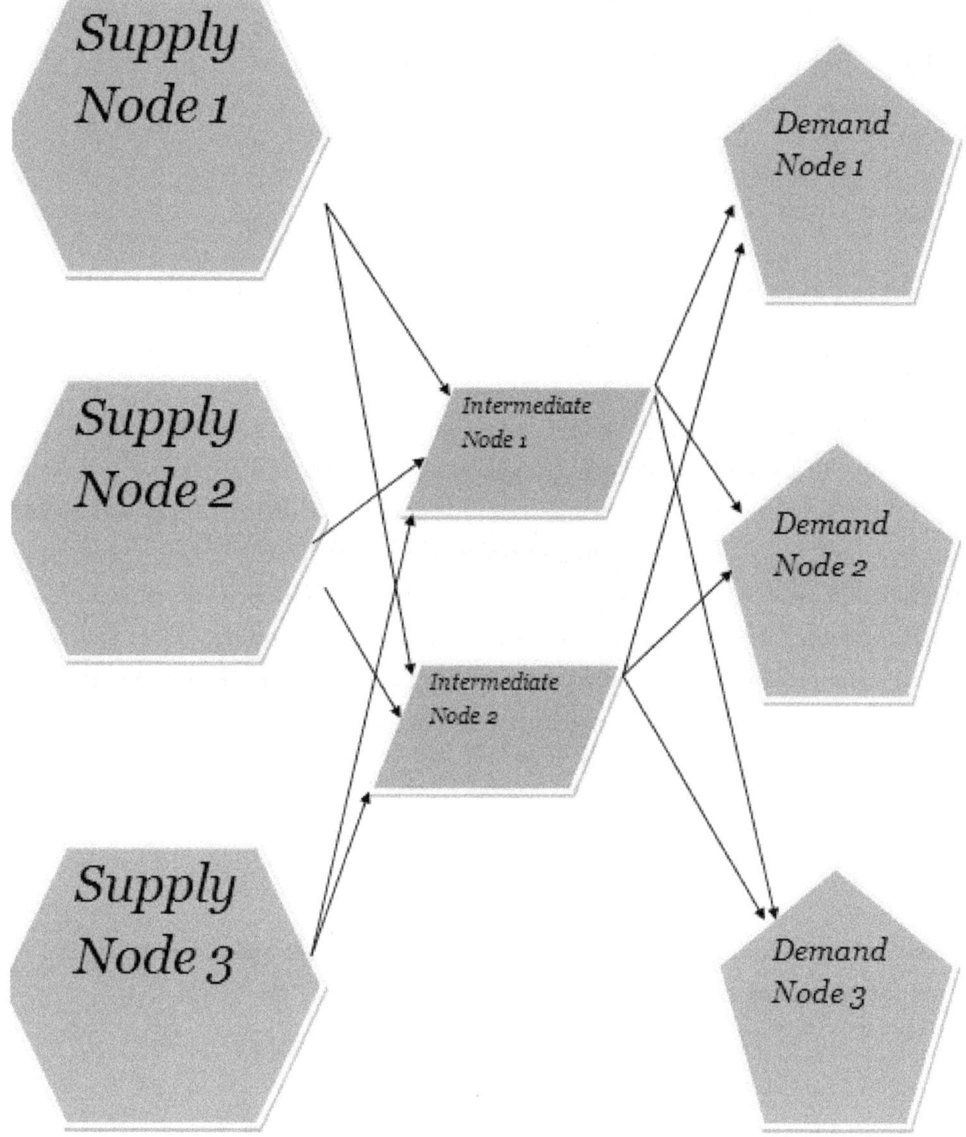

Figure 1: Visual image of supply, intermediate, and demand nodes for transshipment problem

some widgets to NY or Chicago and then fly them to their final destinations. The cost of flying a widget are shown next. Widgetco wants to minimize the total cost of shipping the required widgets to customers. [15]

	NY	Chicago	LA	Boston	Dummy	Supply
Memphis	$8	$13	$25	$28	$0	150
Denver	$15	$12	$26	$25	$0	200

- NY $0 $6 $16 $17 $0   350

- Chicago $6 $0 $14 $16 $0   350

- Demand 350   350   130   130   90

- Supply points: Memphis, Denver

- Demand Points: LA Boston

- Transshipment Points: NY, Chicago [15]

The above example shows a typical transshipment problem. The supply points are Memphis and Denver, capable of producing 150 and 200 widgets respectively. Demand points are Los Angeles and Boston, of which the requirements are 130 widgets each. Note here that the supply and demand amounts do not equal, there for a dummy demand point which will "receive" the excess supply will have to be introduced in order to solve this problem. The intermediate transshipment nodes are New York and Chicago, and the objective is to minimize the total cost of shipping. The data list in this example shows the costs for the shipments between the supply and intermediate nodes as well as the costs between the intermediate nodes and the demand nodes. This example will be solved at a subsequent point in this chapter.

Examples of Transshipment Problems

The manager of Harley's Sand and Gravel Pit has decided to utilize two intermediate nodes as transshipment points for temporary storage of topsoil.

Costs:

From/To	Warehouse 1 (4)	Warehouse 2 (5)
Farm A (1)	3	2
Farm B (2)	4	3
Farm C (3)	2.5	3.5

From/To	Project 1 (6)	Project 2 (7)	Project 3 (8)
Warehouse 1 (4)	2	1	4
Warehouse 2 (5)	3	2	5

[8]

Supply and Demand for this example will be assumed to be the following:

Supply

Farm A: 300 units

Farm B: 400 units

Farm C: 150 units

Demand

Project 1: 200 units

Project 2: 250 units

Project 3: 400 units

Again, the objective in this example is to minimize the costs. The difference in this example, however, is that the supply and demand are equal, therefore, there is no need to introduce a dummy factor into this problem.

8.2 Basics of Linear Programming

Linear programming is defined as, "a mathematical method for determining a way to achieve the best outcome in a given mathematical model for some list of requirements represented as linear relationships." [3] In other words, this is a technique used to optimize a linear objective function, which in turn is subject to linear equality constraints, thus in the outcome, producing the optimal minimal or maximum solution.

For this type of problem, there can be any given number of variables to be determined, along with the known coefficients for these variables.

Let **x = variable to be determined**

Let **c = cost coefficient**

The objective function of the linear programming problem would then be:

$$(\min/\max)\ f(x_1, x_2.....) = c_1 x_1 + c_2 x_2 + c_3 x_3....$$

The outcome for the problem will be subject to the problem constraints as called for by the gathered data, e.g. (supply and demand, cost, quantity, time, etc.)

Constraints should reflect the limits of the data set and should be of the form:

$$x_{11} + x_{12} + x_{13}.... <= limit$$

$$x_{21} + x_{22} + x_{23}... <= limit$$

$$x_{12} + x_{22} + x_{32}... >= limit$$

In addition, we will assume that the variables are always non-negative.

To further illustrate this topic, we will look at an example:

Linear Programming Example

Joe's Diamond shop needs to ship supplies of diamonds to two customers who have made recent orders. Megabucks has ordered 75 custom made diamonds, and Poor Man's Jewelry has ordered 100 custom made diamonds. Joe has two plants that can supply these diamonds for his customers. Sweet Diamond plant is capable of producing 50 diamonds and Sour Diamonds plant is capable of producing 125 diamonds. The costs for shipping the diamonds are as follows:

Sweet Diamonds to Megabucks 6

Sour Diamonds to Megabucks   8

Sweet Diamonds to Poor Man's 5

Sour Diamonds to Poor Man's 9

Joe wants to find a way to ship the diamonds that will cost the least.

To solve this problem, we will label the following nodes:

Sweet Diamonds - node 1

Sour Diamonds - node 2

Megabucks - node 3

Poor Man's Jewelry - node 4

We want to minimize the total shipping cost, so the objective function will be:

min $6x_{13} + 5x_{14} + 8x_{23} + 9x_{24}$

The decision variables are the possible routes between each supply and demand point, and the cost coefficients are the costs associated with each route.

Next, we have to formulate the constraints for this equation. The constraint formulas are based on the supply capacities and the demands of the customers. These are the following:

$$x_{13} + x_{14} \leq 50$$

$$x_{23} + x_{24} \leq 125$$

$$x_{13} + x_{23} \geq 75$$

$$x_{14} + x_{24} \geq 100$$

And again, we will assume non- negativity.

Solving this problem can be done in a variety of ways. Much software is available that can easily solve linear programming problems, thus saving the solver a tremendous amount of time. This example has only a few decision variables, but in reality, problems could have multiple, even up to hundreds of thousands of decision variables. Probably the most widely used software for simple linear programming problems is the Microsoft Office Excel Solver.   Some other software types that can be used are:

- LINDO   http://www.lindo.com/

- AIMMS   http://www.aimms.com/operations-research/mathematical-programming/linear-programming

- GNU Linear Programming kit   http://www.gnu.org/software/glpk/

- PHP Simplex   http://www.phpsimplex.com/en

- Optimalon   http://www.optimalon.com/

8.3 Setting Up the Transshipment Problem

We will now examine the transshipment problem and solve the problem using linear programming.

The formulation of the transshipment problem is as follows:

We will let x_{ab} represent the shipment from node a to node b

We will let y = supply origin and z = destination p = intermediate node

Optimization equation is:

$$\min \sum \sum_{ab} c_{ab} x_{ab}$$

Constraints will be:

$$\sum_{b} x_{ab} <= y_a$$

$$\sum_{a} x_{ap} - \sum_{b} x_{pz} = 0$$

$$\sum_{a} x_{ab} >= z_b$$

$$x_{ab} >= 0 \text{ (non-negativity)}$$

In this formulation, the optimization equation considers the cost as the shipment moves from point a to point b. The supply constraint says that we must leave the supply node and go to

another node. The constraint for the intermediate node says that if we come to this node, we must leave the node and move to another node. Finally, the demand constraint says that this node must be reached from a connecting node.

Example I

Now, we will revisit the previous examples mentioned earlier in this chapter.

Recall:

Widgetco manufactures widgets at two factories, one in Memphis and one in Denver. The Memphis factory can produce as 150 widgets, and the Denver factory can produce as many as 200 widgets per day. Widgets are shipped by air to customers in LA and Boston. The customers in each city require 130 widgets per day. Because of the deregulation of airfares, Widgetco believes that it may be cheaper first fly some widgets to NY or Chicago and then fly them to their final destinations. The cost of flying a widget are shown next. Widgetco wants to minimize the total cost of shipping the required widgets to customers. [15]

	NY	Chicago	LA	Boston	Dummy	Supply
Memphis	$8	$13	$25	$28	$0	150
Denver	$15	$12	$26	$25	$0	200
NY	$0	$6	$16	$17	$0	350
Chicago	$6	$0	$14	$16	$0	350
Demand	350	350	130	130	90	

- Supply points: Memphis, Denver

- Demand Points: LA Boston

- Transshipment Points: NY, Chicago [15]

\

Figure 2 shows a pictorial image of the nodes for our first example problem. Our objective is to find the optimal shipping solution for this problem. We will set this problem up according to the procedure previously shown, then we will solve the problem using the Excel Solver.

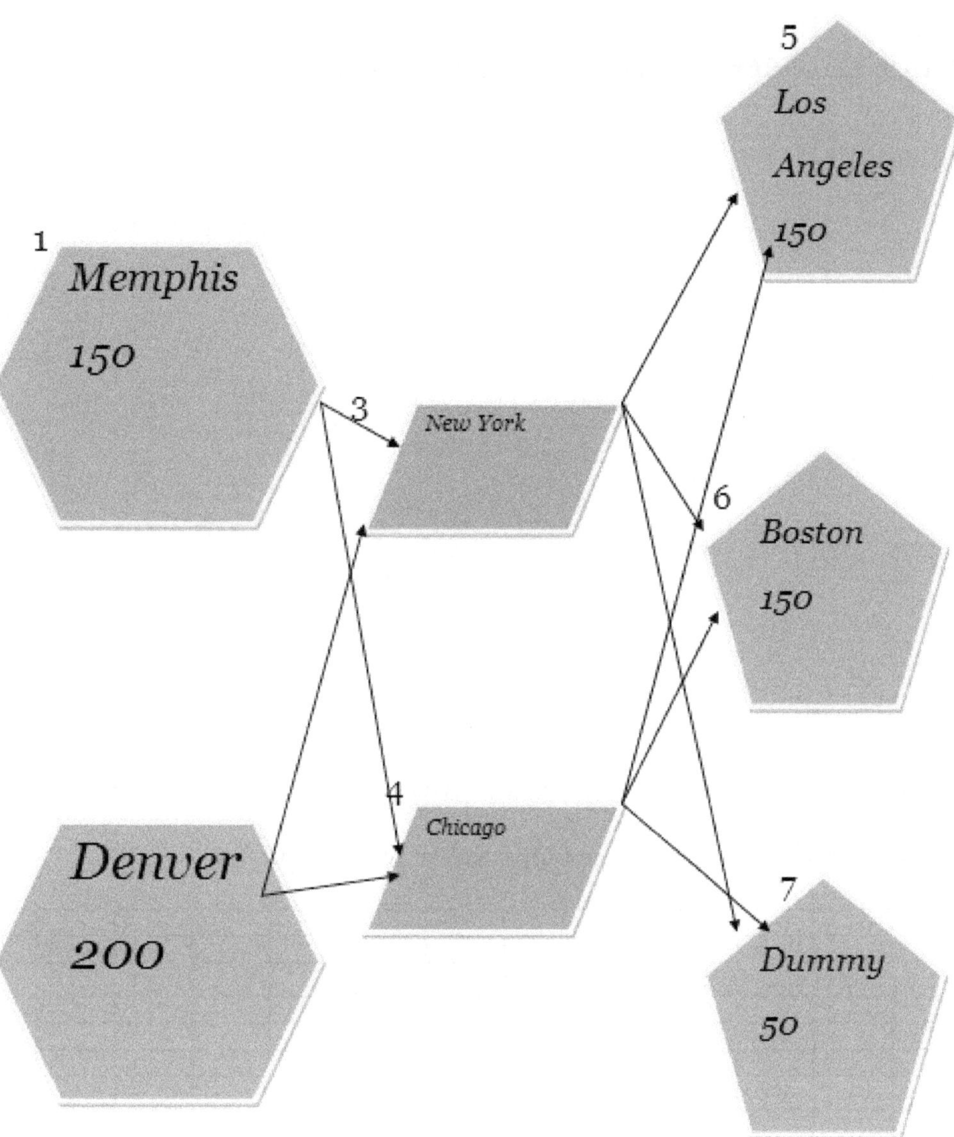

Figure 3: Pictorial Image of Nodes for Example 1

Optimization equation is as follows:

$$\text{Min } 8x_{13} + 13x_{14} + 15x_{23} + 12x_{24} + 16x_{35} + 17x_{36} + 0x_{37} + 14x_{45} + 16x_{46} + 0x_{47}$$

Subject to:

$$x_{13} + x_{14} <= 150$$

$$x_{23} + x_{24} <= 200$$

$$x_{13} + x_{23} - x_{35} - x_{36} - x_{37} = 0$$

$$x_{14} + x_{24} - x_{45} - x_{46} - x_{47} = 0$$

$$x_{35} + x_{45} >= 150$$

$$x_{36} + x_{46} >= 150$$

$$x_{37} + x_{47} >= 50$$

Inputting this data into the excel solver and assuming non-negativity gives the following results:

Table I: Excel solver results showing the optimal shipping solution for Example I

Min $8x_{13} + 13x_{14} + 15x_{23} + 12x_{24} + 16x_{35} + 17x_{36} + 0x_{37} + 14x_{45} + 16x_{46} + 0x_{47}$

S.T.

$x_{13} + x_{14} <= 150$

$x_{23} + x_{24} <= 200$

$x_{13} + x_{23} - x_{35} - x_{36} - x_{37} = 0$

$x_{14} + x_{24} - x_{45} - x46 - x_{47} = 0$

$x_{35} + x_{45} >= 150$

$x_{36} + x_{46} >= 150$

$x_{37} + x_{47} >= 50$

Decision Variables	x13	x14	x23	x24	x35	x36	x37	x45	x46	x47	SUM	RHS limit
Change values	150	0	0	200	0	100	50	150	50	0		
Objective to minimize	8	13	15	12	16	17	0	14	16	0	8200	
Constraint 1	1	1	0	0	0	0	0	0	0	0	150	150
Constraint 2	0	0	1	1	0	0	0	0	0	0	200	200
Constraint 3	1	0	1	0	-1	-1	-1	0	0	0	0	0
Constraint 4	0	1	0	1	0	0	0	-1	-1	-1	0	0
Constraint 5	0	0	0	0	1	0	0	1	0	0	150	150
Constraint 6	0	0	0	0	0	1	0	0	1	0	150	150
Constraint 7	0	0	0	0	0	0	1	0	0	1	50	50

From the above results, we can see that the total minimum cost for shipping is $8200, which entails the following shipping solution:

150 units will be shipped from Memphis to New York

0 units will be shipped from

0 units will be shipped from

200 units will be shipped from Denver to Chicago

0 units will be shipped from New York to Los Angeles

100 units will be shipped from New York to Boston

50 units will be shipped from New York to Dummy

150 units will be shipped from Chicago to Los Angeles

50 units will be shipped from Chicago to Boston

0 units will be shipped from Chicago to Dummy

8.4 Example 2

We will now revisit Example 2 as discussed in previous sections.

Recall:

The manager of Harley's Sand and Gravel Pit has decided to utilize two intermediate nodes as transshipment points for temporary storage of topsoil.

Costs:

From/To	Warehouse 1 (4)	Warehouse 2 (5)
Farm A (1)	3	2
Farm B (2)	4	3
Farm C (3)	2.5	3.5

From/To	Project 1 (6)	Project 2 (7)	Project 3 (8)
Warehouse 1 (4)	2	1	4
Warehouse 2 (5)	3	2	5

[8]

Supply and Demand for this example will be assumed to be the following:

Supply

Farm A: 300 units

Farm B: 400 units

Farm C: 150 units

Demand

Project 1: 200 units

Project 2: 250 units

Project 3: 400 units

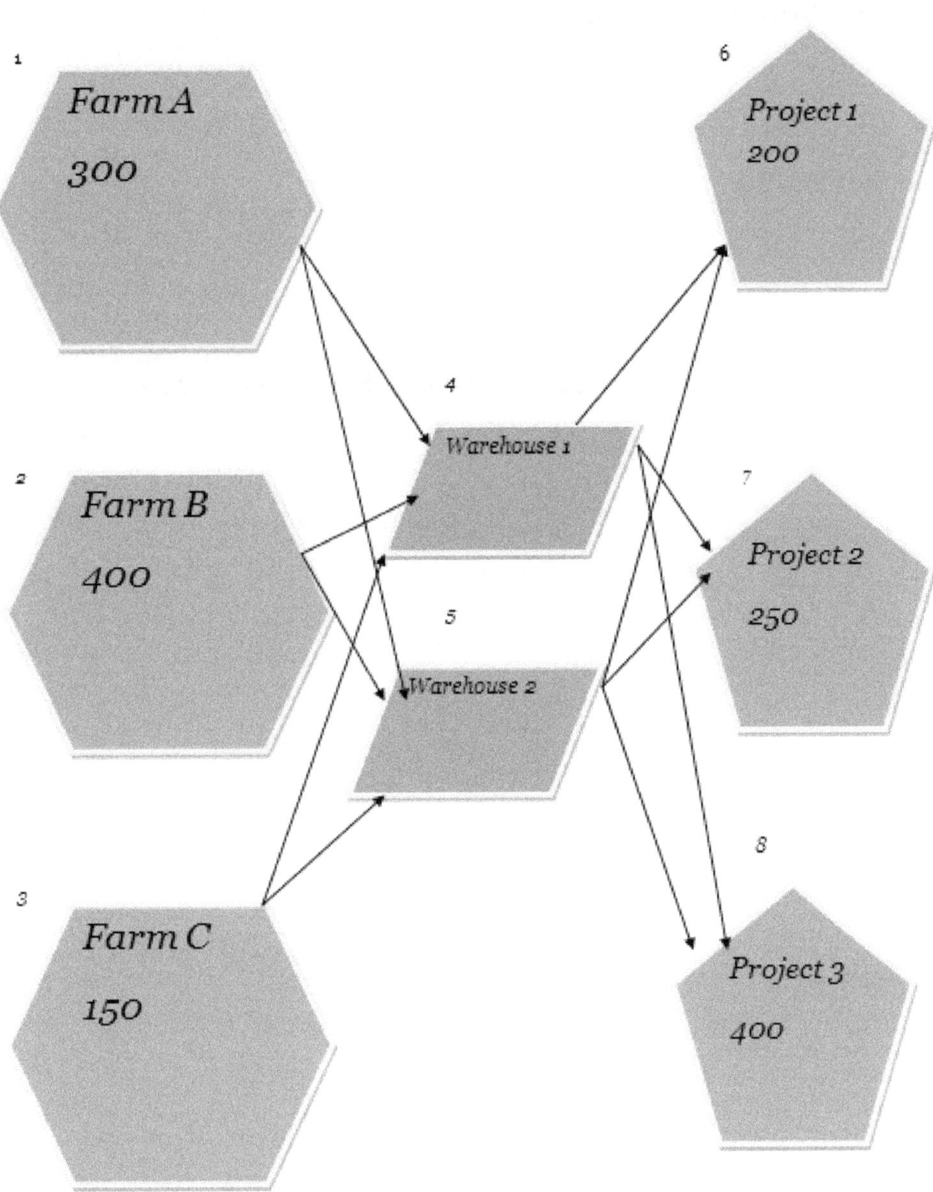

Figure 3: Pictorial Image of Supply, Intermediate, and Demand Nodes for Example 2

Figure 3 shows a pictorial image of the nodes for our second example problem. Our objective again is to find the optimal shipping solution for this problem. We will set this problem up according to the procedure previously shown, then we will solve the problem using the Excel Solver. The difference between this example and the previous example is that both the supply and demand totals for this problem are equal, therefore, no dummy factor need be included to solve this problem.

Optimization equation is the following:

$$\text{Min } 3x_{14} + 2x_{15} + 4x_{24} + 3x_{25} + 2.5x_{34} + 3.5x_{35} + 2x_{46} + 1x_{47} + 4x_{48} + 3x_{56} + 2x_{57} + 5x_{58}$$

Subject to:

$$x_{14} + x_{15} <= 300$$

$$x_{24} + x_{25} <= 400$$

$$x_{34} + x_{35} <= 150$$

$$x_{14} + x_{24} + x_{34} - x_{46} - x_{47} - x_{48} = 0$$

$$x_{15} + x_{25} + x_{35} - x_{56} - x_{57} - x_{58} = 0$$

$$x_{46} + x_{56} >= 200$$

$$x_{47} + x_{57} >= 250$$

$$x_{48} + x_{58} >= 400$$

Inputting data into the Excel Solver and assuming non-negativity gives the following results:

Table 2: Excel solver results showing the optimal shipping solution for Example 2

Optimization Equation		x14	x15	x24	x25	x34	x35	x46	x47	x48	x56	x57	x58		SUM		RHS limit
Min $3x_{14} + 2x_{15} + 4x_{24} + 3x_{25} + 2.5x_{34} + 3.5x_{35} + 2x_{46} + 1x_{47} + 4x_{48} + 3x_{56} + 2x_{57} + 5x_{58}$																	
S.T.																	
$x_{14} + x_{15} <= 300$																	
$x_{24} + x_{25} <= 400$																	
$x_{34} + x_{35} <= 150$																	
$x_{14} + x_{24} + x_{34} - x_{46} - x_{47} - x_{48} = 0$																	
$x_{15} + x_{25} - x_{35} - x_{56} - x_{57} - x_{58} = 0$																	
$x_{46} + x_{56} >= 200$																	
$x_{47} + x_{57} >= 250$																	
$x_{48} + x_{58} >= 400$																	
Decision Variables		x14	x15	x24	x25	x34	x35	x46	x47	x48	x56	x57	x58		SUM		RHS limit
Change values		0	300	300	100	150	0	100	150	200	100	100	200				
Objective to minimize		3	2	4	3	3	4	2	1	4	3	2	5		5125		
Constraint 1		1	1	0	0	0	0	0	0	0	0	0	0		300		300
Constraint 2		0	0	1	1	0	0	0	0	0	0	0	0		400		400
Constraint 3		0	0	0	0	1	1	0	0	0	0	0	0		150		150
Constraint 4		1	0	1	0	1	0	-1	-1	-1	0	0	0		0		0
Constraint 5		0	1	0	1	0	1	0	0	0	-1	-1	-1		0		0
Constraint 6		0	0	0	0	0	0	1	0	0	1	0	0		200		200
Constraint 7		0	0	0	0	0	0	0	1	0	0	1	0		250		250
Constraint 8		0	0	0	0	0	0	0	0	1	0	0	1		400		400

From the above results shown in Table 2, we can see that the optimal shipping cost for this problem is $5125, which entails the following shipping solution:

0 units will be shipped from Farm A to Warehouse 1

300 units will be shipped from Farm A to Warehouse 2

300 units will be shipped from Farm B to Warehouse 1

100 units will be shipped from Farm B to Warehouse 2

150 units will be shipped from Farm C to Warehouse 1

0 units will be shipped from Farm C to Warehouse 2

100 units will be shipped Warehouse 1 to Project 1

150 units will be shipped from Warehouse 1 to Project 2

200 units will be shipped from Warehouse 1 to Project 3

100 units will be shipped from Warehouse 2 to Project 1

100 units will be shipped from Warehouse 2 to Project 2

200 units will be shipped from Warehouse 2 to Project 3

8.5 Concluding Remarks

As stated previously, the overall objective of a transshipment problem is to minimize the costs for this special case of a transportation problem. In today's business world, plans are developed with a long standing goal of saving money and maximizing profit. Many times, companies will find it quicker and easier to send a direct shipment from the supply to the demand location, not realizing that there are other routing opportunities available that will save money. There are also many other factors that must be considered when developing an optimal shipping route. Tradeoffs must be considered when using a transshipment problem for your shipping plans.

For example, it may be quicker for a company to use intermediate shipping nodes to get their product to the demand location, but it may take additional time. If the customer order is urgent, using the intermediate node may not be the optimal solution, even though it may be the most cost efficient. Also, lead time must be taken into consideration if the intermediate node is to be used. If your company has a promised lead time, the additional time it takes to use the intermediate node must be taken into consideration before the transshipment route can be taken. Good customer service is the key to success in any business, which includes minimizing the transportation cost, but also includes assuring the products will be at the demand location at the promised time. Key to making sure all of these things happen is good design of a company supply chain. Transshipment is just one of the many tools available to supply chain managers to help make their company a success.

References

1. "Constraint (mathematics)" - Retrieved from - http://en.wikipedia.org/wiki/Constraint_%28mathematics%29 - Retrieved on May 5, 2013.

2. "Demand" - retrieved from - http://en.wikipedia.org/wiki/Demand - Retrieved on May 5, 2013.

3. "*Linear Programming*" - *Retrieved from* - http://en.wikipedia.org/wiki/Linear_programming. Retrieved on May 6, 2013

4. "Mathematical Optimization" - Retrieved from - http://en.wikipedia.org/wiki/Mathematical_optimization - Retrieved on May 5, 2013.

5. "Node(networking)" - Retrieved from http://en.wikipedia.org/wiki/Node_%28networking%29 - Retrieved on May 5, 2013.

6. "Supply (economics)" - Retrieved from - - http://en.wikipedia.org/wiki/Supply_%28economics%29- Retrieved on May 5, 2013.

7. "Transport hub" - http://en.wikipedia.org/wiki/Transport_hub - Retrieved on May 5, 2013.

8. *"Transportation, Transshipment, and Assignment Problems." 2007. McGraw-Hill/Irwin. Retrieved from* - highered.mcgraw-hill.com/sites/dl/free/007299066x/.../so1e_chap06.ppt. Retrieved on March 17, 2013

9. "Transshipment Problem - Retrieved from - http://en.wikipedia.org/wiki/Transshipment_problem Retrieved on May 6, 2013.

10. AIMMS, The Modeling Company. - Retrieved from - http://www.aimms.com/operations-research/mathematical-programming/linear-programming Retrieved on May 6, 2013.

11. GLPK. - Retrieved from - http://www.gnu.org/software/glpk/. Retrieved on May 6, 2013.

Lecture 9

Power Point Handouts

Logistics/Supply Chain Customer Service

Customer Service in Planning Triangle

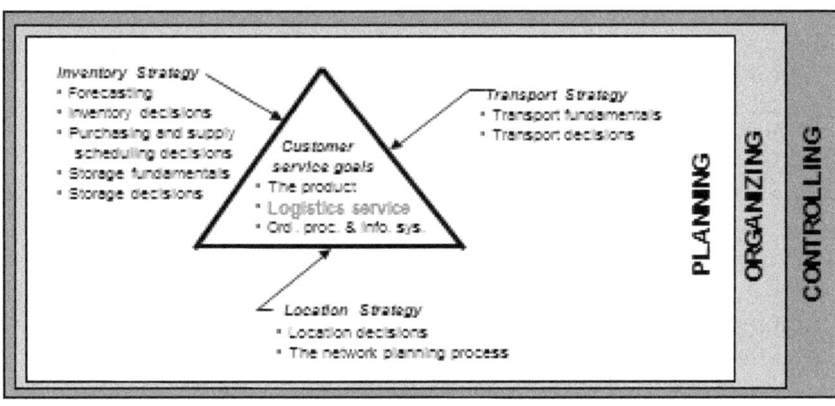

Customer Service Defined

- Customer service is generally presumed to be a means by which companies attempt to differentiate their product, keep customers loyal, increase sales, and improve profits.

- Its elements are:
  - Price
  - Product quality
  - Service

- It is an integral part of the marketing mix of:
  - Price
  - Product
  - Promotion
  - Physical Distribution

- Relative importance of service elements
  - Physical distribution variables dominate price, product, and promotional considerations as customer service considerations
  - *Product availability* and *order cycle time* are dominant physical distribution variables

Customer Service Elements

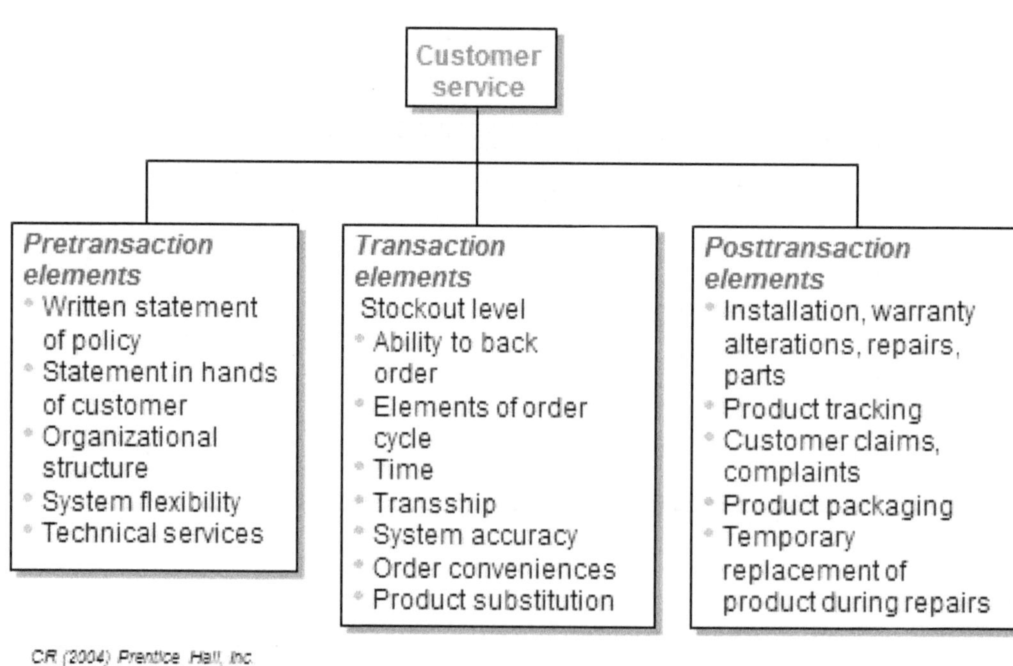

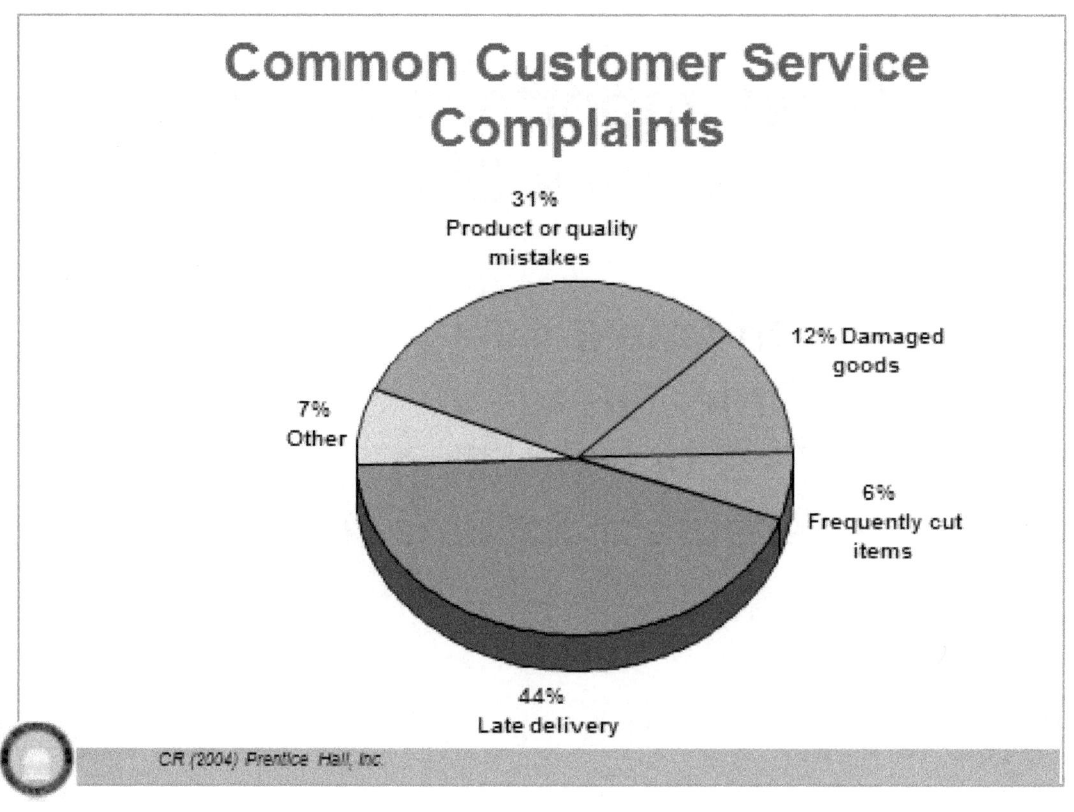

Common Customer Service Complaints
31% Product or quality mistakes
12% Damaged goods
7% Other
6% Frequently cut items
44% Late delivery
CR (2004) Prentice Hall, Inc.

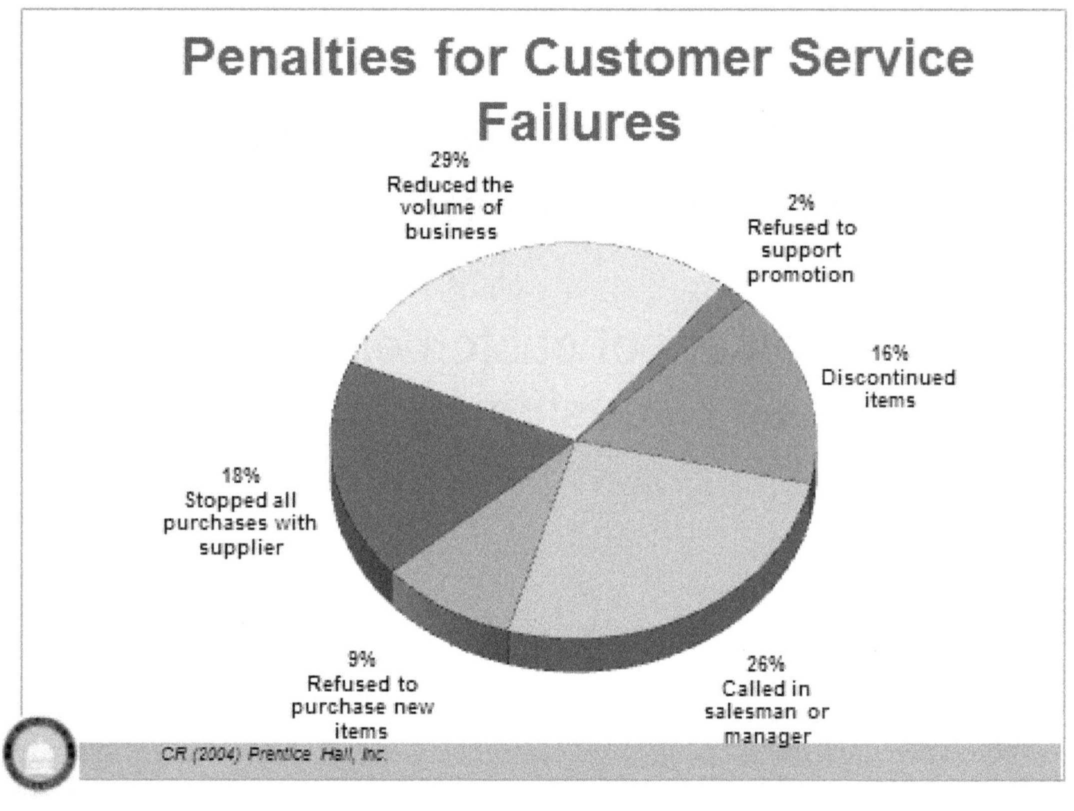

Penalties for Customer Service Failures
29% Reduced the volume of business
2% Refused to support promotion
16% Discontinued items
18% Stopped all purchases with supplier
9% Refused to purchase new items
26% Called in salesman or manager
CR (2004) Prentice Hall, Inc.

Most Important Customer Service Elements

- On-time delivery
- Order fill rate
- Product condition
- Accurate documentation

Appraise This Measure of Logistics Customer Service

Percent of customer orders shipped by customer request date

Parker-Hannifin Corp.

Order Cycle Time

- Order cycle time contains the basic elements of customer service where logistics customer service is defined as:

  the time elapsed between when a customer order, purchase order, or service request is placed by a customer and when it is received by that customer.

- Order cycle elements
    - Transport time
    - Order transmittal time
    - Order processing and assembly time
    - Production time
    - Stock availability

- Order cycle time is expressed as a bimodal frequency distribution

- Constraints on order cycle time
    - Order processing priorities
    - Order condition standards (e.g., damage and filling accuracy)
    - Order constraints (e.g., size minimum and placement schedule)

Components of a Customer Order Cycle

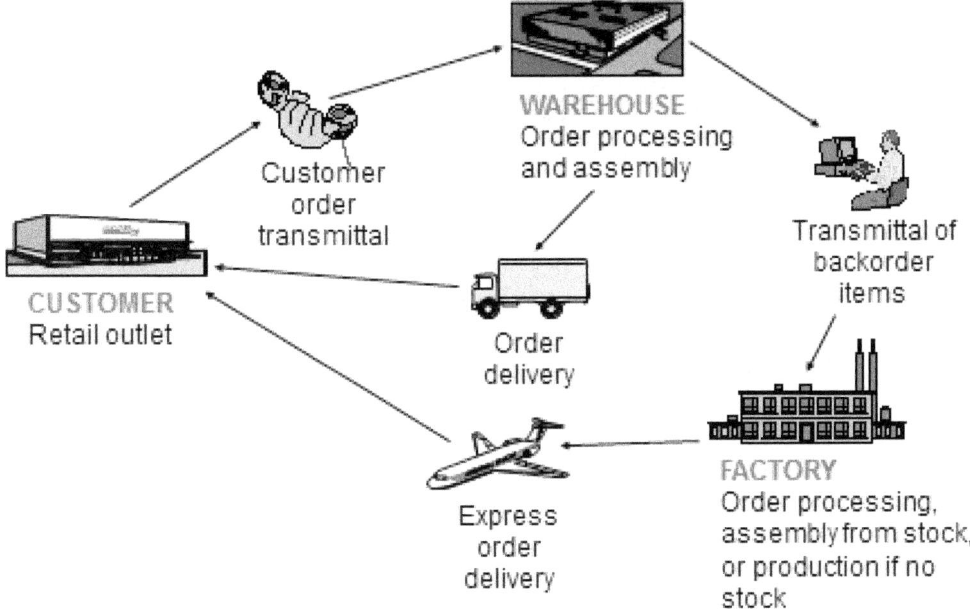

Importance of Logistics Customer Service

- Service affects sales
  - From a GTE/Sylvania study:

    ...distribution, when it provides the proper levels of service to meet customer needs, can lead directly to increased sales, increased market share, and ultimately to increased profit contribution and growth.

  - Service differences have been shown to account for 5 to 6% variation in supplier sales

- Service affects customer patronage
  - Service plays a critical role in maintaining the customer base:

    On the average it is approximately 6 times more expensive to develop a new customer than it is to keep a current one.

Service Observations

- The dominant customer service elements are logistical in nature

- Late delivery is the most common service complaint and speed of delivery is the most important service element

- The penalty for service failure is primarily reduced patronage, i.e., lost sales

- The logistics customer service effect on sales is difficult to determine

Modeling a Sales-Service Relationship

- A mathematical expression of the level of service provided and the revenue generated

- It is needed to find the optimal service level

- A theoretical basis for the relationship

- Methods for determining the curve in practice
    - Two-points method
    - Before-after experiments
    - Game playing
    - Buyer surveys

Remember *Revenue* in ROLA

Sales-Service Relationship by the Two-Points Method

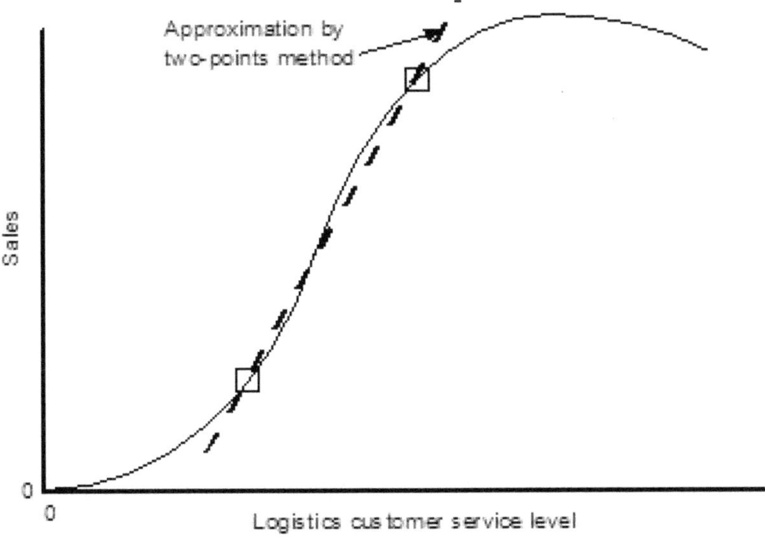

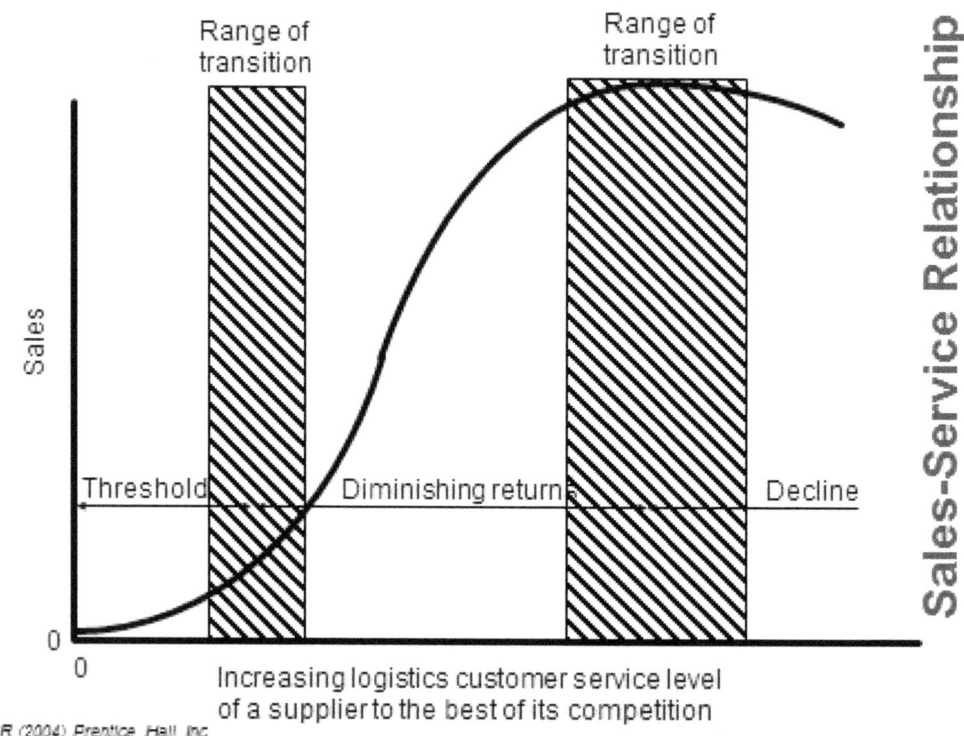

CR (2004) Prentice Hall, Inc.

Determining Optimum Service Levels

- Cost vs. service
- Theory
  - Optimum profit is the point where profit contribution equals marginal cost
- Practice
  - For a constant rate,
    - ΔP = trading margin × sales response rate × annual sales
    - ΔC = annual carrying cost × standard product cost × demand standard deviation over replenishment lead-time × Δz
  - Set $\Delta P = \Delta C$ and find Δz corresponding to a specific service level

CR (2004) Prentice Hall, Inc.

Generalized Cost-Revenue Tradeoffs

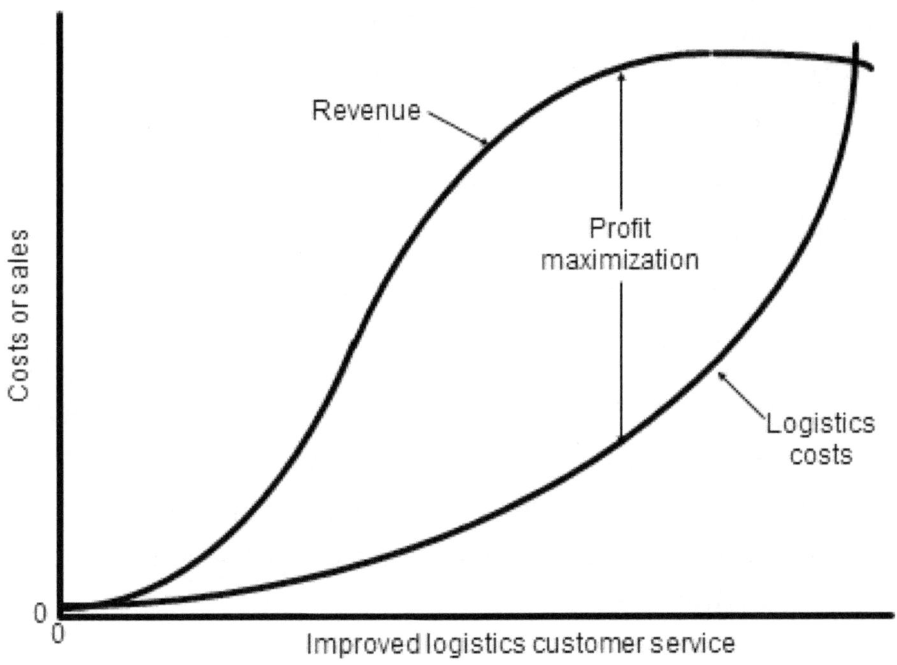

Determining Optimum Service Levels
(Cont'd)

- Example
  - Given the following data for a particular product

    Sales response rate = 0.15% change in revenue
    for a 1% change in the
    service level (fill rate)
    Trading margin = $0.75 per case
    Carrying cost = 25% per year
    Annual sales through the warehouse = 80,000 cases
    Standard product cost = $10.00
    Demand standard deviation = 500 cases over LT
    Lead time = 1 week

Determining Optimum Service Levels
(Cont'd)

Find ΔP

$$\Delta P = 0.75 \times 0.0015 \times 80{,}000$$
$$= \$90.00 \text{ per year}$$

Find ΔC

$$\Delta C = 0.25 \times 10.00 \times 500 \times \Delta z$$
$$= 1250\,\Delta z$$

Set $\Delta P = \Delta C$ and solve for Δz, i.e., $90.00/1250 = \Delta z$

$$\Delta z = 0.072$$

For the change in z found in a normal distribution table, the optimal in-stock probability during the lead time (SL^*) is about 92%.

ΔSL Levels in % for Various Δz Values

ΔSL (%) U L	z_U	$-$	z_L	$=$	Δz
87–86	1.125	–	1.08	=	0.045
88–87	1.17	–	1.125	=	0.045
89–88	1.23	–	1.17	=	0.05
90–89	1.28	–	1.23	=	0.05
91–90	1.34	–	1.28	=	0.06
92–91	1.41	–	1.34	=	0.07 ⇐
93–92	1.48	–	1.41	=	0.07 ⇐
94–93	1.55	–	1.48	=	0.07 ⇐
95–94	1.65	–	1.55	=	0.10
96–95	1.75	–	1.65	=	0.10
97–96	1.88	–	1.75	=	0.13
98–97	2.05	–	1.88	=	0.17
99–98	2.33	–	2.05	=	0.28

*Developed from entries in a normal distribution table

Graphically Setting the Service Level

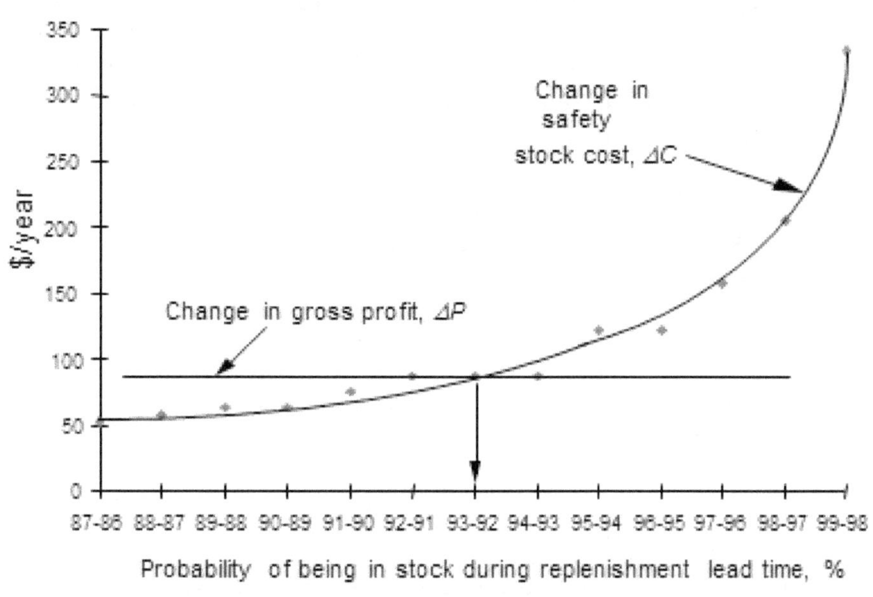

Optimizing on Service Performance Variability

Setting service variability according to Taguchi

- A loss function of the form $L = k(y - m)^2$
  L = loss in \$
  k = a constant to be determined
  y = value of the service variable
  m = the target value of the service variable

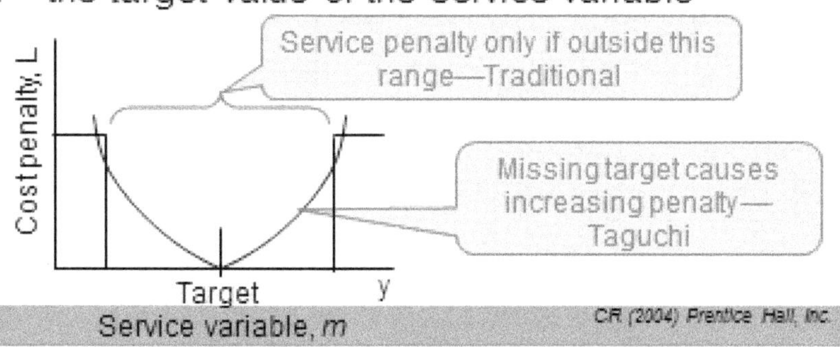

Optimizing on Service Performance Variability (Cont'd)

Setting the allowable deviation from the target service level m is to optimize the sum of penalty cost for not meeting the service target and the cost of producing the service.

$$TC = \text{service penalty cost} + \text{service delivery cost}$$

If the service delivery cost is of the general form $DC = A - B(y-m)$, then find the optimum allowed deviation from the service target.

$$TC = k(y - m)^2 + A - B(y - m)$$

$$\frac{dTC}{d(y - m)} = 2k(y - m) + 0 - B = 0$$

$$y - m = \frac{B}{2k}$$

Marginal delivery cost = marginal penalty cost

If m is set to 0, y is the optimal deviation allowed from target

---

Service Variability Example

Example Pizzas are to be delivered in 30 minutes (target.) Pizzas delivered more than 10 minutes late incur a penalty of $3 off the pizza bill. Delivery costs are estimated at $2, but *decline* at the rate of $0.15 for each minute deviation from target. How much variation should be allowed in the delivery service?

Find k

$$L = k(y - m)^2$$
$$3 = k(10 - 0)^2$$
$$k = \frac{3}{10^2} = 0.03$$

Convert fixed penalty to Taguchi-style loss curve

and y if m is taken as 0

$$y - 0 = \frac{0.15}{2(0.03)} = 2.5 \text{ minutes}$$

No more than 2.5 minutes should be allowed from the 30-minute delivery target to minimize cost.

Setting Service Levels

- Service treated as a constraint on design
- Planning for service contingencies

Measuring Service Performance

- Percent of sales on backorder
- No. of stockouts
- Percent of on-time deliveries
- No. of inaccurate orders
- Order cycle time —— Most comprehensive
- Fill rate--% of demand met, % of orders filled complete, etc.

Service Contingencies

System Breakdown Actions

- Insure the risk
- Plan for alternate supply sources
- Arrange alternate transportation
- Shift demand
- Build quick response to demand shifts
- Set inventories for disruptions

Product Recall Actions

- Establish a task force committee
- Trace the product
- Design a reverse logistics channel

LECTURE 9

READING MATERIALS

9.0 Introduction

Customer service is a broad term that holds many elements ranging from product availability to after-sale maintenance. Logistics perspective, customer service is the outcome of all logistics activities or supply chain processes, and design of the logistics system to be offered.

9.1 Customer Services Defined

Logistics customer service is a part of a firm's overall service offering, the service from a firm's perspective and then distill out those elements that are specific to logistic. The term fulfillment process has been described as the entire process of filling the customer's order. The process includes the receipt of the order, managing the payment, picking and packing the goods, shipping the package, delivering the package, providing customer service for the end user and handling the possible return of the goods. The definitions and descriptions of customer service are broad and need further refinement if we are to use them effectively.

9.1.1 Elements of Customer Services

Marketing has often been described in terms of an activity mix of four P's – product, price, promotion, and place, where place best represents physical distribution. A study of customer service, by the National Council of Physical Distribution Management, has identified the elements of customer service according to when the transaction between the supplier and customer take place.

- Pre-transaction elements establish a climate for good customer service.
    - Written statement of policy
    - Statement in hands of customer
    - Organizational structure
    - System flexibility
    - Technical services
- Transaction elements are those that directly result in the delivery of the product to the customer.
    - Stock out level
    - Ability to back-order
    - Elements of order cycle
    - Time
    - Transship
    - System accuracy
    - Order conveniences
    - Product substitution
- Post-transaction elements represent the array of services needed to support the product in the field; to protect consumers from defective products; to provide for the return of packages; and to handle claims, complaints, and returns.
    - Installation, warranty, alterations, repairs, parts
    - Product tracking
    - Customer claims, complaints

- Product packaging
- Temporary replacement of product during repairs

Corporate customer service is the sum of all these elements because customers react to the total mix.

9.2.2 Relatives Importance of Service Elements

Sterling and Lambert studied the office systems and furniture industry and the plastic industry in some depth. A large number of variables representing product, price, promotion, and physical distribution, they were able to determine those that were most important to the buyer, customers, and influencers of purchases from these industries. Sterling and Lambert research certainly suggests that logistics customer service is dominant in the minds of customers in the office systems and furniture industry and plastic industry. A high fill rates, frequency of delivery, and information on inventory availability, projected shipping date, and projected delivery date at the time of order placement received high rating among the retail customer base. The following are considered the most important logistics customer service elements:

- On-time delivery
- Order fill rate
- Product condition
- Accurate documentation

9.3 Order Cycle Time

Order cycle time can be defined as the elapsed time between when a customer order, purchase order, or service request is placed and when the product or service is received by the customer. Order cycle contains all the time related events that make up the total time required for a customer to receive an order. The order transmittal time may be composed of several time elements, depending on the method used for communicating order. Another major component of order cycle time is for order processing and assembly. Order processing involves such activities as preparing shipping documents, updating inventory records, coordinating credit clearance, checking the order for errors, communicating with customers and interested parties within the company on the status of orders, and disseminating order information to sales, production, and accounting. Order assembly includes the time required to make the shipment ready for delivery after the order has been received and the order information has been made available to the warehouse or shipping department. To a degree, order processing and assembly take place concurrently, so the total time expended for both activities is not the sum of the times required by each. Stock availability has a dramatic effect on total order cycle time because it often forces product and information flows to move out of the established channel. The final primary element in the order cycle over which the logistician has direct control is the delivery time, the time required to move the order from the stocking point to the customer location.

9.3.1 Adjustments to Order Cycle Time

It has been assumed that the elements of the order cycle have been operating without constraint but customer service policies will distort the normal order cycle time patterns.

9.3.2 Order Processing Priorities

Distinguishing one customer from another may be necessary when backlogs occur. An individual customer may vary greatly from the company standard, depending on the priority rules, or lack of them, that have been established for processing incoming orders.

9.3.3 Order Condition Standards

What is a normal order cycle time can be substantially altered if the products ordered arrive at the customer location damaged or unusable state. Standards set for package design, procedures for returning and replacing incorrect or damaged goods, and standards set for monitoring order quality will establish how much the order cycle time will be increased on the average.

9.3.4 Order Constraints

The logistician may find it desirable to impose a minimum order size, to have order placed according to a preset schedule, or to have order forms prepared by the customer that conform to preset specifications. These constraints permit important economies to be achieved in product distribution. On the other hand, this practice may allow service to be provided to some low-volume markets that might not otherwise be served very frequently or reliably.

9.4 Importance of Logistics / SC Customer Service

Logistics managers may be tempted to dismiss customer service as a marketing or sales department responsibility. But the key concern at this point is weather it makes a difference to the selling firm in any way that can affect its profitability. How service affects sales and how service affects customer loyalty is questions that need to be explored.

9.4.1 Service Effects on Sales

Logisticians have long believed that sales are affected to some degree by the level of logistics customer service provided. The fact is that logistics customer service represents an element within total customer service, sales cannot be precisely measured against the levels logistics customer service, and buyers themselves do not always accurately express their desires for service and consistently respond to service offerings. There is definitive evidence that logistics customer service does affect sales. Sterling and Lambert did a study on customer service, they were able to conclude that marketing services do affect market share and that the marketing mix components of product, price, promotion, and physical distribution do not contribute equally to market share. But Baritz and Zissman were able to show that customers can perceive service differences among their best and their average suppliers. They also observed that when service failures occur, buyers often impose penalizing action on the responsible supplier. Singhal and Hendricks did a study of 861 publicly held companies found that supply chain failures have an adverse effect on stock prices. The six most common reasons for supply chain glitches were: parts shortages, changes requested by customers, new product ramp / rollouts, production problems, development problems, and quality problems.

9.4.2 Service Effects on Customer Patronage

To look at the importance of customer service is through the costs associated with customer patronage. Logistics customer service plays a critical role in maintaining customer patronage and must be carefully set and consistently provided if customers are to remain loyal to their supplier.

9.5 Defining a Sales-Service Relationship

It is not clear how important is logistics customer service. But logistics decision making would be enhanced if we knew more precisely how sales change with changes in logistics customer service levels. When no customer service exists between a buyer and a supplier, or when service is extremely poor, little or no sales are generated. As service increased to that approximating the offering by competition, little sales gain can be expected. When a firm's level reaches this threshold, further service improvement relative to competition can show good sales stimulation. It is possible that service improvements can be carried too far, with a resulting decline sales.

9.6. Modeling the Sales-Service Relationship

The sales service relationship for a given product may deviate from the theoretical relationship. Many methods for modeling the actual relationship might be used in these specific cases.

9.6.1 Two-Points Method

The two point's method involves establishing two points on the diminishing return portion of the sales service relationship through which a straight line can be drawn. The method is based on the notion that multiple data points to accurately define the sales service curve would be expensive or unrealistic to obtain, and if data were available, it is not usually possible to describe the relationship with a great deal of accuracy. First set logistics customer service at a high level for a particular product and observing the sales that can be achieved. Then the level is reduced to a low level and sales are gain noted. These limitations suggest that a careful selection of the situation to which it is to be applied must be made if reasonable results are to be obtained.

9.6.2 Before-After Experiments

The sales response to a particular change in service may be all that is needed to evaluate the effect on costs. The sales service curve over a wide range of service choices may be unnecessary and impractical. Before and after experiments of this type are subject to the same methodological problems as the two point's method.

9.6.3 Game Playing

One problem in measuring the sales response to service changes is controlling the business environment so that only the effect of the logistics customer service level is determined. One approach is to set up a laboratory simulation, or gaming situation, where the participants make their decisions within a controlled environment. The artificiality of the gaming environment will always lead to question about the relevance of the results to a particular firm or product situation.

9.6.4 Buyer Surveys

The popular method for gathering customer service information is to survey buyers or other persons who influence purchases. Mail questionnaires and personal interviews are frequently used because a large sample of information can be obtained at relatively low cost. Survey methods must be used with caution because biases can occur. The question must be carefully designed so as not to lead the respondents or to bias their answers and yet capture the essence of service that the buyers find important.

9.7 Costs versus Service

The age old question of most successful logistics programs are where they choose to be in relationship to cost and service levels. Saving money and keeping freight spends down are paramount to the success of many operations. In the demanding global economy, the quality of service must be considered to achieve the complexities of the modern logistics cycle. The perfect mix of service and cost are the mystery that eludes most logistics professionals. Determining the optimum cost with the optimum service level is critical and essential to operation balance.

9.8 Determining Optimum Service Levels

9.8.1 Theory

Identifying the revenue and cost for each service level will provide the logistics professionals a starting point to make this critical decision. Taking into consideration the demands of his organization and customers they can then decide on a direction utilizing the pricing information. Revenue, cycle time, and inventory costs are only some of the factors to determine the optimum solution.

9.8.2 Practice

Net profit is the driving force in many logistics scenarios. The optimum service level is found when the net profit is maximized in the operation. $NP = P\text{-}C$ is a simple formula to arrive at the maximum net profit. P is the gross profit and C is the safety stock cost at the warehouse. The fluctuation of the safety stock usually determines the service level for the warehouse.

Although net profit in a warehouse is essential, determining logistics decisions about transportation has many factors and one key factor is quality. A shipment arriving on time in the condition intended is a key factor in customer service. Imagine you have ordered you child a stereo for Christmas over the internet. The package should arrive December 22nd at your home in plenty of time for wrapping and you pleasantly pleased with the free shipping offered. The package leaves on time and you are tracking it to your home in anticipation. Now it is Christmas Eve and you do not have your package and your unhappiness is growing with every second. The package arrives December 278th and looks like it was dropped from the truck on the way. Yes, your transportation costs expectations were meat but your quality was far from expected. The mix of the two is the ideal spot for customer service and happiness.

9.9 Service Variability

Variability is a powerful term in the logistics customer service arena. The global economy has contributed greatly to the variability in customer service. Instead of depending on a supplier from Illinois to deliver a component that product now comes from the other side of the planet. Ocean ship's capacities and schedules have thwarted many a great plan and reputation. How much variability can be tolerated by a customer is the million dollar question?

9.9.1 Loss Function

The expectations of customers usually are measured by the meeting of deadlines and schedules by logistics systems. Missed deadlines adversely affect other schedules that adversely affect other schedules and so forth. Genichi Taguchi developed a formula for loss function that is defined as $L=k(y\text{-}m)^2$. L= loss in dollars per unit, y= value of the quality variable, m= target value of the quality variable, and k= a constant that depends on the financial importance of the quality variable. When the loss function is known a company or logistics team can set the tolerance of the variable of the customer service level.

9.9.2 Information Substitution

Substituting information to customers for the variability of service levels is a very intelligent modern approach. Sometimes you don't know what you don't know is a dangerous outlook on schedules and deliveries. Knowing what you don't know by tracking information and high visibility allows the customer to adjust to the variability quicker with less impact on their operations. Emails, texts and mobile communication can allow customers to see everything going on and successfully communicate the variables allowing better customer service.

The internet retail operation is taking unbelievable customer service improvements by utilizing high visibility of an order. Emails sent at every step allow the company to set expectations as they go for each order. Apple and Dell have perfected this visibility on their products ordered online as you see every step of the operation and feel more involved with the process as you receive your computer or Apple product. High visibility sets these companies up for high customer service levels with just keeping their customers informed.

9.10 Service as a Constraint

Customer service can be a constraint to a logistics system. Service levels set by competitors and often traditional service levels can affect the customer service and cost relationship. Sensitivity analysis can help aid a logistics operation to determining the factors that constrain the operation. The ideal solution is still the optimum balance between quality and cost; this should be weighed heavily in all analysis of the constraint of the operation.

9.11 Measuring Service

A great defining section of the customer service process is measuring service. This is a very difficult process in logistics as many elements exist that affect the outcome. There are many performance measures in logistics; order documentation accuracy, order entry, transportation, product availability, product damage in transport and processing time. Understanding your customer's expectations is important to determine which factors or measures are important to high customer service. Not understanding your customer's expectations can allow a competitor to identify the shortcoming and capitalize on the factor. An example of this would be a manufacturer of snowmobiles gets a new Canadian customer that gives them an order for 100 snowmobiles. The manufacturer ships mostly to the Northern US where roads are paved and smooth. Now their product is going to Northern Canada where roads are rough and not paved. The manufacturer makes no packaging changes and the snowmobiles arrive in Northern Canada where they are met with a parade as the locals desperately need them. The snowmobiles show up to the parade damaged and covered in mud and muck completely unusable. Simple Shrink wrapping or tarping and additional packaging could have solved this problem but this customer service demand was never identified.

9.12 Service Contingencies

Most of the time a logistics operation runs smoothly and as planned. There is times where troubleshooting is needed to overcome short term issues. Product recall or system breakdown are two issues of many that can arise to demand contingency plans. Natural disasters and terrorist acts are also issues that will demand quick thinking and solution resolution from logistics professionals.

9.12.1 System Breakdown

A global economy has inherently a very complex logistical system. Getting a raw material from China to the US manufacturer and then the final product back to Japan can have many factors that can cause a system breakdown. Weather, natural disaster, economic upheaval or even political changes can affect a logistical system in many drastic ways. Inventory is the attribute of a supply chain or logistical system that will allow them to strive in one of these dramatic events. Inventory will allow the system the time it needs to recover to pre event performance levels.

9.12.2 Product Recall

Product recalls are becoming more and more the norm of businesses today. This drastic reverse logistics scenario can wreak havoc on any logistical system. Designing and effective reverse distribution process for the recall is essential for the survival of a company. Proper contingency planning for this scenario will allow a company to flip the switch for this process and minimize the impact on the health of the company.

9.13. Concluding Comments

Customer service is a very important measure of the efficiency of a logistical system. Many measures and processes allow the logistics professional an opportunity to receive feedback from the customer on their efficiency. The term that the customer is always right may not always be true but certainly reigns supreme in most companies. The complexity added by a global economy has increased the visibility of the customer service in logistics and emphasizes the importance of measuring and examining the process. Customer service will influence many decisions in logistics and require much analysis for optimum performance.

Lecture 10

Power Point Handouts

Transportation Rates

Transport Fundamentals in Planning Triangle

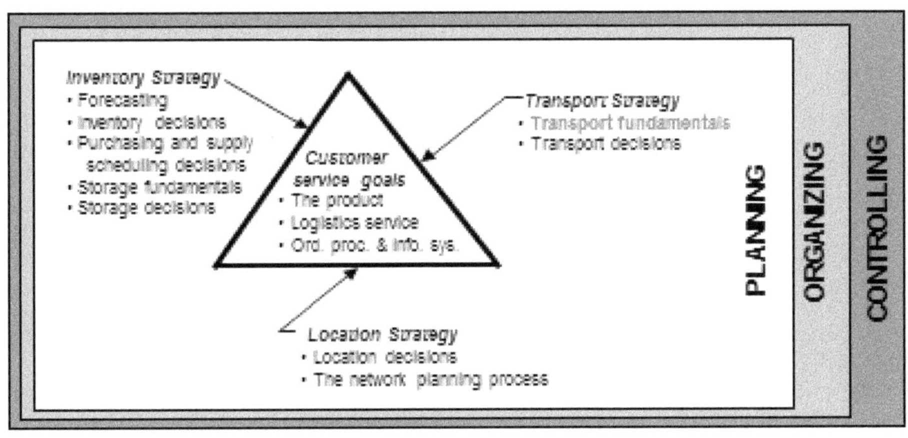

Transport System Defined

- *Performance*
  - Average transit time
  - Transit time variability
  - Loss and damage
  - Other factors including availability, capability, frequency of movement, and various less tangible services

- *Cost*
  - Line haul
  - Terminal/local
  - Accessorial or special charges

Transport Choices

- *Primary intercity carriers*
  - Air
  - Truck
  - Rail
  - Water
  - Pipe

- *Coordinated services*
  - Piggyback
  - Birdyback
  - Fishyback

- *Small shipment carriers*
  - UPS
  - Federal Express
  - Postal services
  - Bus Package Express

- *Agents*
  - Freight forwarders
  - Shipper associations

- *Others*
  - Autos
  - Bicycles
  - Taxis
  - Human
  - Electronic

Importance of Modes

By Products Hauled

- **Air**--very high-valued, time sensitive products

- **Truck**--moderately high-valued, time sensitive products. Many finished and semifinished goods

- **Rail**--low-valued products including many raw materials

- **Water**--very low-valued products moved domestically, high-valued if moved internationally

- **Pipe**--generally limited to petroleum products and natural gas

Importance of Modes (Cont'd)

By Volume Moved

Transportation mode	Percent of total volume
Railroads	36.5%
Trucks	24.9
Inland waterways	16.3
Oil pipelines	22.0
Air	0.3
Total	100.0

Performance Overview

- Air generally fast over long distances and a fair degree of relative variability
- Water is very slow and moderately reliable
- Pipe is very slow but reliable
- Truck is moderately fast and reliable
- Rail is slower and less reliable than truck

Relative Costs of Performance

Mode	Price, ¢/ton-mile
Rail	2.28
Truck	26.19
Water	0.74
Pipeline	1.46
Air	61.20

Legal Classification

- Common carriers
- Contract carriers
- Private
- Agents

Documentation

- Bill of lading
- Freight bill
- Freight claims

International Transportation

- Free trade zones
- Documentation
- Modes

Foreign (Free) Trade Zone

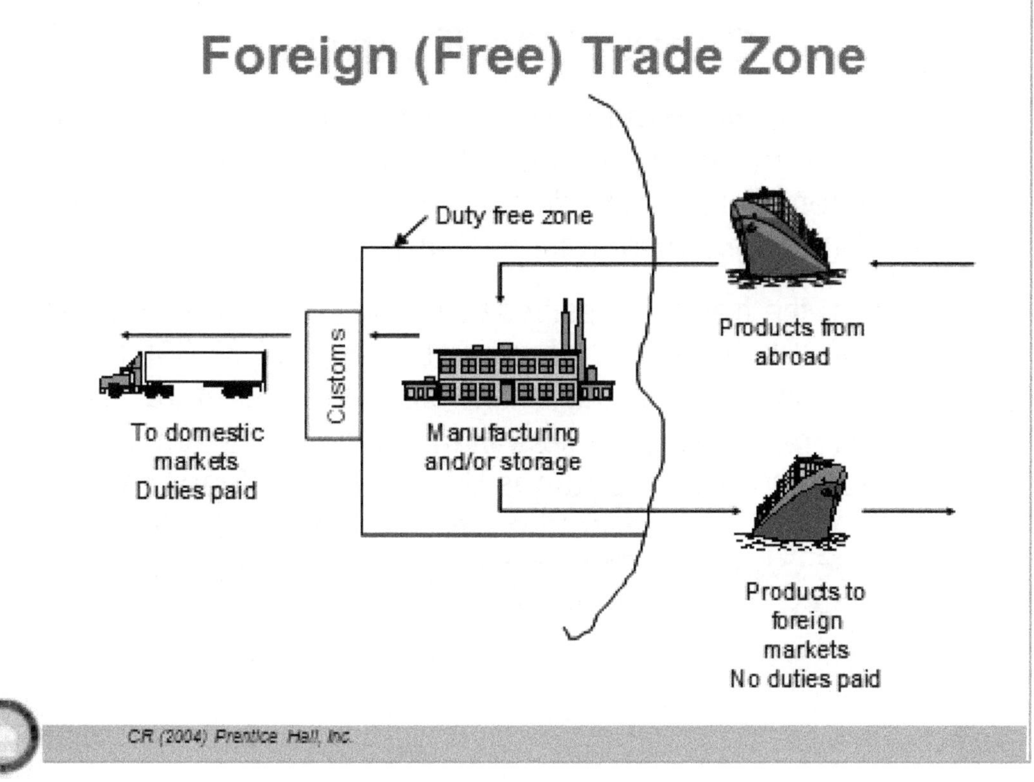

Rate Types

- *Line haul rates*
  - Class
    - >Freight classification of items
    - >Rate tables of tariffs
  - Contract rates
  - Drayage (local delivery)
- *Commodity and contract rates*
  - Specific rates for given shipment sizes for specific products moving between designated points
- *Special service charges*
  - Extra charges
  - Stop-off privilege example
- *Private carrier costing*

Class Rate Example

Suppose we wish to ship 15,000 lb. (150 cwt.) of wheat flour from New York to Los Angeles by truck. The trucker offers a 40% discount from the published tariff. What is the transportation charge?

From the freight classification table, this is item number 1090--00. It shows a minimum weight of 36,000 lb., which is less than this shipment size. Therefore, the class rating is 55, or less-than-truckload. From the class 100 tariff, the rate is 6065, or $60.65 per cwt. With a 40% discount, the effective rate is (1- .40) x 60.65 = $36.39. The shipment charges are 0.55 x 36.39 x 150 = **$3,002.18**.

CR (2004) Prentice Hall, Inc.

Break Weight

Question Suppose 9,000 lb. of Class 100 merchandise is to be shipped from New York to Dallas. From Table 6-4, the rate would be $52.21/cwt. However, should the shipment be priced at the next higher weight break rate of $40.11/cwt. for a lower cost?

$$Break\ Weight = \frac{Rate_{Next} \times Weight_{Next}}{Rate_{Current}}$$

where:

$Break\ Weight$ = Weight above which the next higher weight break rate should be used for lower transport costs

$Rate_{Next}$ = Rate for next higher weight break

$Weight_{Next}$ = Minimum weight of next higher weight break

$Rate_{Current}$ = Rate for true weight of shipment.

CR (2004) Prentice Hall, Inc.

Break Weight (Cont'd)

Answer

Calculate break weight

$$\text{Break Weight} = \frac{40.11 \times 100}{52.21} = 76.82, \text{ or } 7{,}682\,\text{lb.}$$

Since the 9,000 lb. shipment size exceeds the break weight of 7,682 lb., size as if a 10,000 lb., shipment for a total cost of $40.11 \times 100 = \$4,011$. Otherwise, the shipment would have cost $\$52.21 \times 90 = \$4,699$.

Stop-Off Privilege Example

Suppose 3 shipments of J=8,000 lb., K=12,000 lb., and L=10,000 lb. originating at I are to be delivered in the following way.

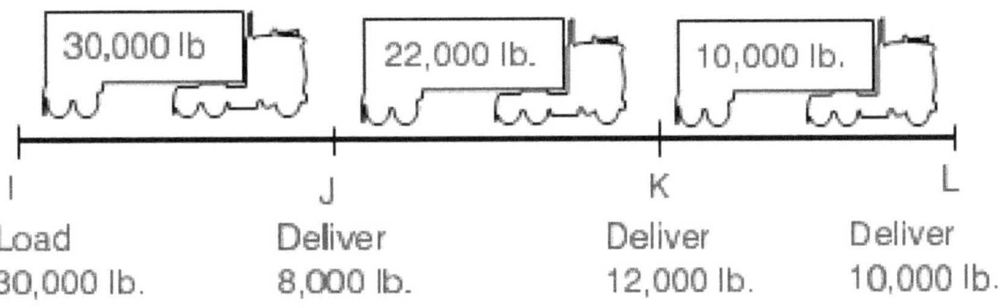

Stop-Off Privilege Example (Cont'd)

First, we compare the costs without the stop-off privilege. This would be to price as if each shipment is a separate shipment from *I*. Suppose we know the rates. Hence,

Load, lb.	Points	Rate, $/cwt.	Charges
8,000	I to J	3.05	$244.00
12,000	I to K	3.35	402.00
10,000	I to L	3.60	360.00
		Total	$1006.00

Now, we price with the stop-off privilege. We assume that all the volume (30,000 lb.) is to be delivered to the farthest stop and we use the rate to that point ($3.00/cwt.). A small stop off charge of $15.00 is made for each stop including the last stop. Hence,

Load, lb.	Points	Rate, $/cwt.	Charges
30,000	I to L	3.00	$900.00
		3 stops at $15 each	45.00
		Total	$945.00

Rate Profiles

By distance

Rates vary with the distance between origin and destination in the following manner

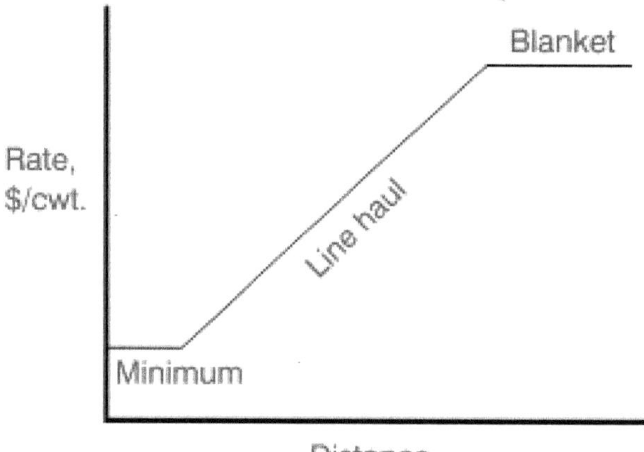

Rate Profiles (Cont'd)

By volume

Rates by shipment size have the following characteristic

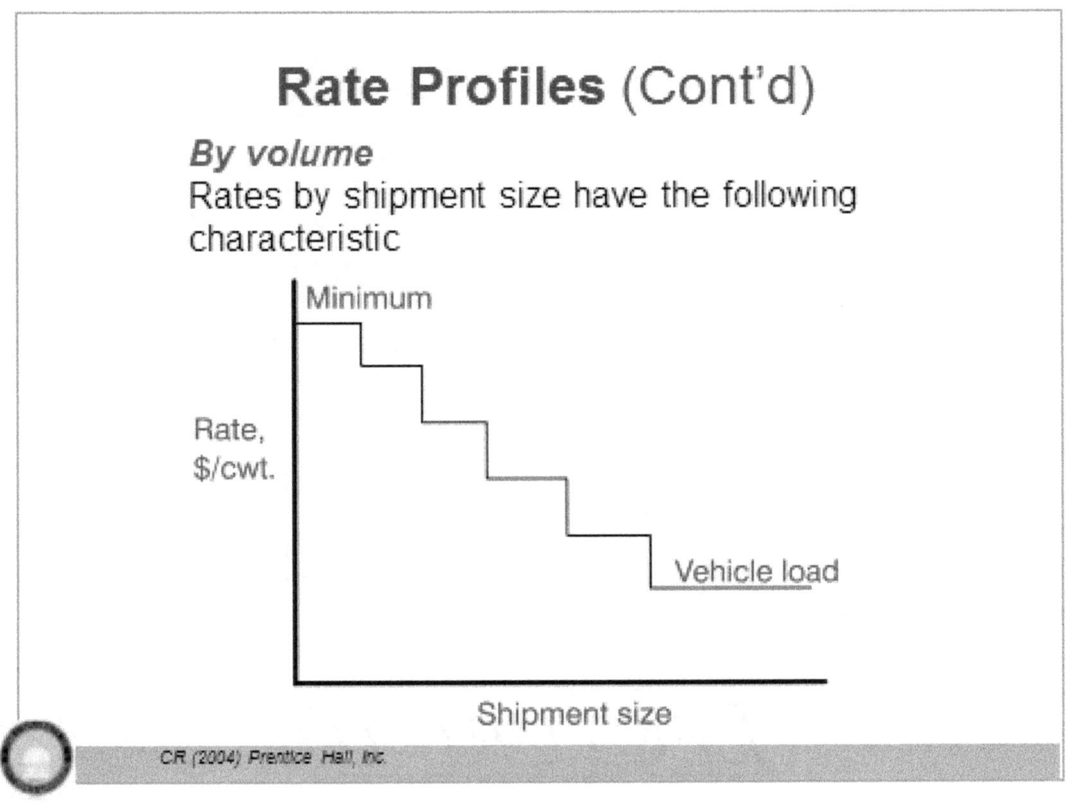

LECTURE 10

READING MATERIALS

10.1 Introduction

Transportation is one of the most important aspects of all basic functions of our society. Not only does transportation allow for the travel of humans from place to place, but also for the transport of products and freight all throughout the world to help the global economy thrive. There are several means of transport, whether transport occurs through the air or across the ocean. For those in the role of logistician, it is important to remember that for every shipment, there is an appropriate transportation that fits best in terms of completing the job and cost savings. This chapter will provide an inside look at what types of transportation are available for certain products, the costs and factors that attribute to those costs for transportation modes, and how a logistician can achieve the lowest possible price for carrier service through negotiation.

10.2 Available Modes of Transportation

Transportation is the means by which people and freight are able to move from their points of origin to many different destinations. Transportation can be provided through a variety of different modes; primarily through three different means (consisting of five different modes of transportation). Transportation can occur on the ground (truck, rail, pipeline), in the air, and through shipping (water). Each of these modes of transportation will be discussed in depth to gain a better understanding of the characteristics, advantages, and disadvantages of each. Figure-1 displays the ranking of the different modes in terms of their cost, delivery time, variability, and loss and damage.

Figure-1 Performance Characteristics

Mode of Transportation	Cost (per ton mile)	Average Delivery Time (door to door speed)	Absolute Delivery-Time Variability	Percent Delivery-Time Variability (ratio of absolute dtv to average dt)	Loss and Damage
	1=Highest	1=Fastest	1=Least	1=Least	1=Least
Air	1	1	1	5	3
Pipeline	4	4	2	1	1
Rail	3	3	4	3	5
Road	2	2	3	2	4
Water	5	5	5	4	2

Air

Air transportation has become a rapidly growing mode of transportation. Air operations may only account for 2% of the global trade measured in weight, but they account for 40% of global trade measured in value [4]. Although air transportation has one of the higher costs and highest levels of variability due to external concerns such as weather, congestion, etc., air is still the most suitable option in many cases because of the unmatched ability to deliver cargo over long distances in a timely manner. Because of this option, air is the chosen mode of transportation for highly perishable freight, emergency supplies, and productions with low inventories. In these cases, the need for supplies to be delivered quickly trumps the cost of the mode of transportation.

Air transportation maintains many routes to almost anywhere domestically and internationally and can provide services that no other mode of transportation can match. It is up to the logistician to determine whether or not the cost of the transportation is worth the amount of time that could be saved in exchange. One problem with air transportation, however, is the amount of cargo that can be carried is confined by the cargo space in the designated aircraft. For example, any large set of cargo with a significant weight would not be suited for air, but perhaps rail or water.

Pipeline

Pipeline transportation may be the least recognizable form of transportation. But, pipelines can be constructed in virtually any environment (underground, in the ocean) over long distances, and they are primarily used to transport gas and liquids (i.e., oil, gas). However, in certain cases, they are also able to transport water in local areas, coal in the form of slurry (dried particles suspended in a liquid), and other solid objects, although these are not as frequent and usually take place in much shorter distances compared to other traditional pipelines. Pipelines are used to connect a natural resource in an isolated area far away to major manufacturing or population centers. Although the delivery of products through pipeline transportation is not very fast, the transportation of products is quite reliable. Because they are built underground and underwater for the most part, they are not subject to external conditions (weather) that many other forms of transportation are exposed to. Pumping operations for pipeline transportation are also very reliable. Another advantage pipeline transportation offers is the significant reduction in loss and damage risk. Once again, because pipelines are primarily carrying liquids and gases, they are not subject to the same type of damage that solid materials may be subject to. Although there are many benefits to utilizing pipeline transportation, the mere fact they are so limited in what they can transport significantly reduces the chances a logistician would be able to use this mode of transportation often.

Rail

Rail transportation is among the middle of the pack in most categories according to Table-1. Perhaps rail's greatest advantage is some of the upgrades they have over road transportation. Rail transportation allows for greater load capacities for their cars, and they can also meet the special needs of the some of their consumers. Their cars may be able to be suited for products requiring refrigeration and other adaptable features to be able to carry other items such as sand, livestock, petrochemical products, fertilizers, and passengers. Unlike road transportation, with each additional container added to the train, the marginal cost decreases until the point where the train's maximum capacity is filled. Trains, however, spend most of their time in between stops, loading and unloading. This feature can be a bonus if the customer needs multiple pickups and deliveries along one route. Another great feature of rail transportation is its versatility. They are able to accept containers directly from trucks, transport them to another destination long off. This method is called "piggybacking," and will be discussed in greater detail in the Intermodal section. Logisticians should especially consider rail as a mode of transportation when the loads to be hauled are large and are the destinations are far enough away that trucking is not efficient.

Road

Road transportation is the form of freight transport that has expanded the most over the last 50 years [4]. Although road transportation has grown, there are several factors that have limited the expansion. Because of governmental restrictions on the allowed size and carrying capacity of trucks for safety purposes, the amount one vehicle can carry is limited. The fixed costs for trucking remains low, but the variable costs take up the significant portion of the cost. These contributing factors may include but are not limited to increase in fuel consumption and prices, traffic congestion causing delays, and maintenance. Road transportation offers several advantages. The greatest of these advantages is the speedy delivery with door to door service. Unlike many other forms of transportation, road transporters do not have to load and unload at ports or terminals. They can drive directly from origin to destination in reasonable amounts of time at low costs. Logisticians should consider road as their primary option if they load sizes are able to reasonably fit within a truckload and for shorter ranges.

Water

Water transportation, or maritime shipping, is the cheapest cost per ton mile for shipping. This is, in large part, because of their low line-haul costs. The only costs they incur include operation of the equipment. Their fixed costs include cost of the equipment and use of the terminal facilities. Water transportation's major advantage is the ability to haul so much cargo at once to so many international ports at low costs. Cargo is typically defined into two categories: bulk cargo, freight that is not packaged, and break-bulk cargo, freight that has been packaged. About 70% of all maritime shipping is bulk cargo. Break-bulk cargo is increasing, however, because of the containerization, reducing the risk of damage and loss.

Water transportation is a primary tool of the industrial sector as it represents the most efficient means of transporting bulk cargo over long distances. Logisticians should consider this mode of transportation if the loads are large and are not heading for locations with low inventory as it is the slowest means of travel.

Intermodal

Intermodal transportation has increased over the years because of the growth of international shipping, and the ease in which modes can be transferred. The primary force that has allowed for intermodal transportation to become prominent in today's world is containerization. Containers to be used for intermodal transportation are generally standard sized metal boxes. As discussed earlier, "piggybacking" is the mode in which loads from truck are transferred to rail. In many cases, the truck container can be removed and loaded directly onto the train to be taken to the next destination. These two modes working together benefits both as truck and rail largely compete for many of the same customers. Another popular form of intermodal transportation is referred to as "fishyback." In this mode, trucks deliver their loads to ships who then deliver the rest of the way. As with "piggybacking," truck containers can be directly loaded onto the ships to allow for easier transfer. There are several other forms of intermodal transportation, but they are not seen as often. Logisticians should consider intermodal transportation when advantages of several modes are able to be seized while maintaining the goal of cost savings and meeting the consumer needs.

10.2.1 Transportation Costs

The costs of transportation are not merely based on the mode of travel. There are inherent costs that are associated with every mode of transportation. This section will specifically focus on the transportation costs that all modes face. The following sections will specifically delve into the rates that affect each individual mode.

The first factor in all costs for each mode will be the transport costs. These are the costs associated with infrastructure (fixed costs) and the costs associated with operating (variable costs). Every type of transportation has these costs but will be different for each one. There are three types of transport costs. Terminal costs- These are the costs associated with the loading and unloading of product. They are essentially unavoidable.

Linehaul costs- These are the costs associated with the function of hauling x amount of freight over a certain amount of distance (fuel, labor).

Capital costs- Costs related to the physical components of the transportation (the actual vehicles, terminals and infrastructure along the way). Physical assets are not forever and can deteriorate. Capital costs are necessary over time, but are not an every trip type of cost.

There are several factors that affect transport costs, each with their own distinct impact.

Geography- geography's impact has to do mainly with the distance traveled because distance is the most basic factor in affect transport costs. Not only the distance covered, but the accessibility of the routes along that distance is a factor as well. This factory can vary immensely depending on the type of mode used.

Product type- The type of product being hauled can affect the transport costs. Perishable products that may require certain storage parameters may increase the costs compared to a non-perishable product.

Economies of scale- As discussed previously, with transportation such as rail, the more units that are being transported, the lower unit cost per unit transported becomes

Energy- All modes of transportation require an energy source for movement, primarily oil. About 60% of all global oil consumption is attributed to transport activities. These costs are highly variable due to the unpredictable shifts in the price of energy.

Infrastructures- the infrastructure of the terminals, ports, etc. may lead to higher costs if they are not efficient.

Competition- Transport costs tend to be higher in markets where there is little competition compared to markets with high competition.

Surcharges- additional fees that may be charged for transport. An example would be baggage fees of the airlines.

Modes- The type of mode transporting the load has significant impacts on how much the total transport costs will be as Figure-2 illustrates. The following section will talk more specifically about each mode.

Figure-2 Average Freight Revenue per Ton Mile

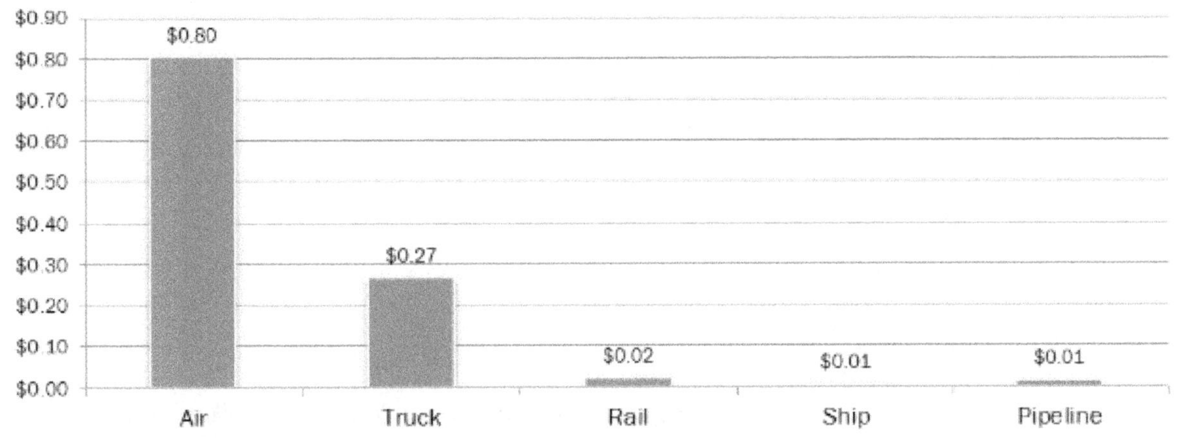

Source: U.S. Department of Transportation, *National Transportation Statistics*, 2009.

Air

High terminal costs with high takeoff costs

High line-haul costs (16x higher than rail)

Fast, but highest cost mode of transport

Becomes more economical with longer distances traveled

Pipeline

Lowest line-haul costs of all modes

High fixed costs that increase with distance covered

Restricted use; regular flow and demand is required

Rail

High terminal costs requiring a large investment

Medium line-haul costs

More effective with an increase in the length of haul

Road

Negligible turnover costs (door to door)

High line-haul costs (4x higher than rail)

Requires high cost for large capacity and long bulk hauls

Water

Terminal costs are the bulk of the cost

Low line-haul costs (1/3 as much as rail)

More effective with the greater the length of haul

Cheapest, but slowest of all modes

10.3 Approaches to Negotiating Transportation Rates

For nearly all modes of transportation, price is negotiable. For those companies who are sticking to fixed rates, they are dinosaurs in terms of keeping up with their competitors. Competition in the transportation market is ever-growing and increasingly demanding due to so many different factors affect the cost of transport (see above). If companies are not willing to negotiate their price, there are competitors who are willing to do so in order to secure business. Two basic ways for negotiating the price for transport are to negotiate with individual carriers and to use an RFP (Request for Proposal), or bid process.

10.3.1 Negotiation with Individual Carriers

Negotiation with individual carriers is the less predominant of the two choices, primarily because it limits the amount of money that could be saved by not using the bid process. However, this choice may be appropriate when the market for transporters for a certain product is narrow. Another reason one company may choose to only negotiate with an individual carrier is that carrier has been employed before exhibiting exemplary service. For example, in this author's experience working with a small office supply chain company, there were several distributors who may have been able to provide their supplies at a lower cost, but the management's relationship with the current distributor was an important factor. He could use the other companies' prices as leverage, but he never had any intention of changing. The same concept can be applied to negotiating with individual carriers.

10.3.2 Using a Bid Process

The more effective of the two options is to use a bid process by submitting an RFP. By using the bid process, logisticians may be able to determine what kind of price the market will yield for the transport of their product by receiving so many returns. Because there is no limit to how many RFPs that can be sent, a greater understanding can be achieved. Since the market for transportation services is so competitive, the bid process can cause competitors to try and seize their opportunities for the business by submitting the lowest possible prices. The advantage to one's business is obvious. For a company who has never used an RFP to solicit proposals from competing firms, the annual savings is estimated to be somewhere in the range of 10-25%.

10.4 Collect and Analyze Data: Preparing an RFP

So, how is an RFP prepared? The bulk of the work load for preparing an RFP is saturated in the collection and analysis of data. The first step is to identify the requirement necessary to be fulfilled for the job. One of the most effective ways is to meet with staff company-wide to obtain a better understanding of the transportation requirements and about current carrier performance (if applicable). The next step would be to provide an in depth analysis of the company's past transportation activities going at least as far back as 6 to 12 months. This report should provide details such as the common origins, destinations, and weights. The total amount of deliveries and averages should also be provided. It is important not to list which companies have previously provided these services in the past, or for what rates those companies provided the service. This will protect the bid process by helping to get the most accurate quotes possible. The next step is to give company growth charts and projections for growth. If transportation companies see that your company is thriving and will be around for the foreseeable future, they will feel more secure in offering a competitive proposal. It is also necessary to have all RFPs uniformly written to provide that all the quotes submitted back are in the same format in regards to pricing. Basically, many companies have different ways for pricing their services, but it is essential that all the companies price in the same way in order to review which offer is the most favorable. After all the information has been gathered and analyzed, a booklet should be prepared listing this information along with the instructions on exactly how the companies should respond to the RFP's with their quote. The final step before sending out the RFP's will be determining which companies to send them to. As stated previously, there is no limit to how many RFP's can be sent out, but it would not make much sense to send them to a company who will be unable to fulfill the required needs. Make sure the companies that receive the RFP will be able to complete the tasks at hand.

When the carrier proposals are finally submitted back, the final decision should be made based not only on cost, but on the stability and capability of the services and any other factor that may be inherently important to the company. Some companies submit RFP's annually to ensure they are getting a fair price for the services they see in return, but if a company chooses not to go this route, it is necessary for them to stay aware of the market trends and be weary of price increases by current carriers.

10.5 Building Strong Service Provider Relationships

Relationships between service providers and customers are important for achieving high levels of customer satisfaction and loyalty, as many professionals and service management scholars have shown. Building from existing theories, the relationships between service providers are another important contributor to customer outcomes. When service processes are highly interdependent, uncertain, and time constrained, relationships between service providers are integral to the process of coordination and therefore are an important contributor to customer outcomes. Strong provider-provider relationships directly increase customer satisfaction and loyalty because the overall service experience is more effectively coordinated. Second, strong provider-provider relationships help transport service providers to develop more effective relationships with their customers, which further increases customer satisfaction and loyalty. Transport managers should therefore select, train, and reward service providers in a way that supports the formation of strong working relationships between them.

The concept of customer service has already pervaded many types of service industries. Accordingly, apart from continuing to emphasize the core benefits that they receive from providing goods and services, companies are placing additional importance on customers' willingness to make repeat purchases. The long-term partnerships, cultivated by means of relationship marketing and customer relationship management, will bring them even greater revenue and profit. If a company strives to provide good customer service, it will consequently wish to make customers aware that the benefit of the service it provides is greater than the sacrifice entailed. Its service can therefore enhance customer value, and is worthy of customer commitment. It is therefore vital for companies, if they are to create long-term relationships benefiting both themselves and their customers. In today's highly competitive operating environment, gaining a full understanding of customers needs and creating new customers is an important part of corporate management. Maintaining the loyalty of existing customers can be a difficult task due to customers' increasingly high service quality demands and the individualized customer needs, most companies lose an average of 25% of their customers every year. Developing a new customer requires roughly five times the cost of maintaining an existing customer. In the wake of globalization, companies must therefore deal with vast amounts of customer data. Hence, the customer relationship management has become a key focus of corporate operations. During the last few years, large international container carriers have steadily entered the international logistics service market. They are relying on investment in subsidiaries under their own brands to establish global shipping carrier based logistics service providers. The global shipping carrier based logistics service providers have been established by large international container carriers in order to create a win-win shipping environment and achieve their transport and logistics goals. As a result, large container carriers have gradually shifted to outsourcing the functions, which has led to the emergence of third party logistics service providers. Container carriers usually rely on alliances involving container communities spanning international logistics chains to create the greatest possible customer value and loyalty, enhance productivity, reduce operating costs and risk, and increase profitability. Customers and carriers are both concerned about whether cargo can be safely transported to its destination. In order to ensure that this goal is reached, cargo logistics effectiveness is especially important to global shipping carrier based logistics service providers. From a marketing perspective, a vital issue is how to enable global shipping carrier based logistics service providers to become efficient logistics service stations creating significant added value for customers, and thereby ensuring full-scale customer success and maintaining the global shipping carrier based logistics service provider's competitiveness.

Businesses within the transportation industry have a much greater amount of exposure to the public than many industries. Consider that many transportation companies not only maintain a customer service department, but have ongoing contact with customers through their drivers, agents, sales representatives and logistics coordinators. Thus, improving customer service in a transportation firm extends far beyond merely improving the customer service department.

Steps involved in improving customer service that may be used by some providers are many, below are few that are commonly used in the industry.

Step 1
Offer customer service training to educate every employee who has direct contact with consumers. Emphasize the importance the company places upon maintaining a high standard of service. At every point of customer contact, employees must recognize customer service as a priority.

Step 2
Communicate the company's expectation for high levels of customer service. Via internal newsletters and emphasize it in management meetings. Implement competitions and contests to generate company specific suggestions on how to enhance customer service. This will not only incentivize the performance of customer service through bottom up participation, but will also encourage active thought regarding continual customer service improvement.

Step 3
Recruit employees with a strong customer focus for public contact positions. Ensure that customer service training is integrated into all new employee training programs. As new hires fill positions due to normal turnover, the company culture will gradually become increasingly customer service oriented.

Step 4

Implement software and support technology to enhance and enable your employees to provide a higher level of customer service. Customer Relationship Management software will allow your employees to enter, access and track customer activity and information. Your staff will have greater access to customer information that will allow them to support the customer service efforts of staff working directly with consumers.

Step 5

Empower your frontline staff to better service customer concerns. Every staff member operating in direct contact with customers should have at least some ability to address their concerns or have a superior address them when they are reported. This might be as simple as issuing mobile phones to drivers or empowering drivers to grant customer credits.

a. Post Negotiation Support

Companies should work on behalf of their customers to obtain the best price, the best terms, and the best service possible; they should realize that building strong relationships with transportation and freight carriers is an important part of the end sum. Services should be designed to help customers get the absolute most out of their service provider relationships, and to ensure that they are equal, financially sound, and beneficial for all parties. Customers should be made to ensure they are represented by ethical, informed, and focused experts, at the rate negotiating game. Negotiation efforts should not stop after negotiations and establishment of the best rates and service. Invoice auditing technology ensures that service providers comply with contractual terms and conditions. Compliance monitoring services, we should also be at the forefront along with, pinpointing overcharges, errors and omissions to ensure contractual satisfaction per the published tariff. A large portion of companies operate in the blind when it comes to negotiating transportation/rule contracts because they leave the negotiating process in the hands of a Traffic Manager, Purchasing Agent or Cost Account who may have little or no understanding of the company's goals beyond cost containment. Those companies that seek to elevate their Transportation Management Department by integrating it into a larger logistics-oriented or leveraged buying group strategy find that unless their transportation negotiators actually worked for an asset based carrier, they only see and hear an outsider's perspective on whether they are getting the best price and service. In negotiating transportation contracts, knowledge is power. The more customers know about your carrier's cost and pricing practices, the better they can negotiate a fair price. Understanding cost drivers from a business unit perspective is only half of freight negotiations the other half is understanding a service providers cost. Some of a major service provider's cost that need to be clearly defined and understood by the shipper are:

- Minutes of down time at origin and destination
- Cubic capacity of shipments
- Handling units
- Load ability
- Density on the run
- Claims ratio
- Break bulk cost
- Lane balance and imbalance
- Miles between stops

- Fixed cost

- Sales personnel

- Negotiating freight costs is an element of the micro logistics component called transportation. Utilizing cost accounting principles and assigning an appropriate and relevant cost will facilitate freight cost negotiations and satisfy the need for meaningful and proper freight rates and charges for your business.

10.6 Common Misconceptions in Rate Negotiation

Many purchasers erroneously continue to take transportation costs as fixed, and therefore not relevant for contract negotiation. There are two primary reasons for this misconception. First, many buyers look at transportation cost as only a small part of the unit price of a specific item. Their attention is focused on the ratio of the transportation rate to the delivered unit price of the item. Often, if the ratio of transportation rate to delivered price is small, the buyer will assume that the transportation cost is insignificant, ignoring the service implications of transportation and the impact that transportation service quality has on inventory and production management. In addition, buyers may evaluate the transportation characteristics of their purchases (volume, frequency, distance, etc.,) and conclude that their traffic is not valued by carriers. Second, in the regulated transportation environment similar carriers in a given mode were required to charge the same price for the same service, and service innovations were sharply limited. There was little or no service or price differentiation between similar carriers in a given mode. The result was that, under regulation, transportation was reduced to a commodity. Many purchasers still hold this view of transportation service. Below are some of those misconceptions of the transportation management industry.

- Freight rate reductions are not linear.

- Discounts do not necessarily mean freight bills are reduced. Accessorial charges, minimum charge thresholds and other factors may eliminate or offset the discount. Still, most shippers use averages and aggregate numbers when analyzing carrier proposals.

- A better rate isn't necessarily a good rate.

- A firm can't know if they are getting competitive freight rates unless they know what other shippers are paying.

- To carriers, all freight is not the same.

- Knowing what freight the carriers want can earn rate reductions. Sales incentives change regularly, so understanding type of freight certain carriers seek is critical. At times splitting freight across multiple carriers to gain the very best rate and service is to a firm's advantage.

- Bundling all your freight with one carrier doesn't ensure the best pricing.

- This is a common misconception with consultants that negotiate freight. Because they think negotiating freight is like negotiating pencils, they miss the nuances of the commodity and leave money on the table.
- You need high volume and leverage to get the best rates.

Many third party logistics companies and brokers will tell a shipper that it is through leverage that they are able to get better rates for their customers. Leverage doesn't drive down rates, but a deep and extensive knowledge of the industry, transportation management, and knowing what can be negotiated, will. Having extensive knowledge, expertise, and experience drives better rates.
Third Party Logistics Companies or brokers will save you money versus what you can do yourself.

Consider how most brokers and 3PLs make their money. They mark up the difference between the rate they negotiate with the carrier and how much they charge. Their goal is to charge as much as possible while creating a perception that they are saving money. This explains why when one broker is competing for business with another broker the saving spread is so minimal.

Most shippers working with brokers and 3PLs have no idea of the real negotiated rate. The broker and 3PL keep that very secret. Shippers are often paying 30, 40, 50% or more than what it actually costs to move a. There is no transparency in the industry as far as this goes.
You will always get the best rates when going directly to the carriers.

If a business opts to work with and negotiate with carriers directly, they need to have extensive industry knowledge as we said before. They need to understand base rate pricing, surcharges, and what the carrier is willing to negotiate. Carriers are experts at creating the perception that a shipper is getting the best rates possible simply by elevating their base rate and then offering deep discounts. Having deep and up-to-date knowledge of the industry can help a business save money when negotiating directly with carriers, however, the cost of handling logistics in-house can be expensive. The company may spend as much money in-sourcing their freight as what they would pay having it handled externally. The fact is carrier pricing is all over the board and their goal is to maximize   the money they can get from a customer for a shipment. There is very little integrity in pricing. A company could literally being paying more than four times as much than their next door neighbor sending a comparable shipment going to the same place.
Fuel Surcharges cannot be negotiated.

Fuel surcharges are a big headline item in the news and industry and get a lot of attention, but fuel surcharges, are a profit item for companies. Fuel Surcharges are negotiable, funds can be saved on fuel surcharges every day. Negotiating prices for each individual customer based on their freight and lanes is the key. Knowledge of the overall process is what drives all pricing attributes, not mythical leverage.
Freight bills are always correct.

Most companies assume that just because a bill came from a carrier or is managed by a 3PL or broker, that the bill must be correct. As a result, most companies don't spend the time to properly review and audit their freight bills. According to industry experience, some freight bills that get issued are incorrect. The errors are generally in favor of the carrier too. In some cases, this could amount to tens of thousands of dollars to a shipper. Auditing and bill matching is essential to ensuring that a company is maximizing the amount of money they save on freight. This of course could get very costly and time-consuming unless a company has the right technology and resources in place to properly audit and review their freight bills for accuracy. In-sourcing is the best way to save money in freight management.
Outsourcing is the best way of saving money in freight management. These are conflicting messages that create a lot of confusion within the industry. As a result companies are often changing from outsourcing to in-sourcing and back depending on the current state of the market place and the internal resources. Sometimes a company goes from handling their freight internally to going with a broker or 3PL because a good sales person convinced them it was in their best interest. In-sourcing can work when a company has

the right systems and people in place with the knowledge and core competency to be effective. This means the company has to keep the right people on board who are producing tangible and cost-effective results. Often businesses overlook some of the hidden costs of in-sourcing. The biggest problem with in-sourcing is the fact that it often distracts companies from focusing on their core competencies. That could cost a company in the way of time, money, energy, and focus. Outsourcing can work when you find the right vendor that can provide quality and consistent results for less than what it would cost to do it yourself. Outsourcing can keep a company focused on its core competency. A poor vendor, however, can cost a company time and money if it provides bad service or doesn't perform as promised. The problem is that a company that outsources can often lose control and perhaps not have access to knowledge, information, and data that would enable a company to make better decisions. When outsourcing, a company is subject to the response times and schedules of the vendor too. One of the things many companies overlook is the cost incurred during the transitions from in sourcing and outsourcing models. These changes can be costly and disruptive to a business as well. A combination of both is the best approach.

10.7 Conclusion

Transportation is an important tool in supply chain management; it is the major connector of the upstream and downstream and is that aspect of the supply chain that takes goods and services to the retail end of the supply chain. It is the major promoter of globalization and as such is one that is to be given high attention to ensure efficiency of the chain. The various modes of transportation are the way in which products and services can be moved from one place to another with each mode having its unique characteristics, its gains and setbacks. The five basic modes of transportation are highway, rail, water, pipeline and air. Their advantages and limitations are as follows. The highway mode has flexibility because items can be delivered to almost any location in a continent. Transit times are good, and rates are usually reasonable for small quantities and over short distances, rail: generally has low cost, but transit times are long and may be subject to variability, water has a very high capacity and very low cost; but transit times are very slow, and large areas of the world are not directly accessible to water carriers, pipeline capabilities are highly specialized and limited to liquids, gases, or solids in slurry form. No packaging is needed, and the costs per mile are low, however the initial cost to build the pipeline is very high. Air transportation is fast but the most expensive. All transportation modes have their pros and cons which make the decision of selecting the ideal transportation mode a difficult one. This is rate negotiation is vital, the key to a successful negotiation is preparation and strategic planning. Utilizing information based on a company's rate analysis, a successful rate negotiation can take place. Generally the two major decisions that need to be made are, trading transportation costs over inventory costs and subsequently overall costs. Trading transportation costs over inventory aggregation. Transportation over inventory costs various factors have to be taken into consideration when choosing the most appropriate transportation mode. Also, the effect of a transportation mode on a supply chain also aids in the decision making process; that is, does one transportation mode make the supply chain more customer responsive than the other even though it is quite expensive, or is there a cheaper mode that responds a lot more slower to customer demands than another. This is where the customer has to make the appropriate decision and it is most likely based on cost. Whatever decision is taken must be a trade-off that best meets the interest of the consumer whilst retaining profitability of the firm. The design task is to trade-off these characteristics to best meet the demands of the marketplace other factors include the ability to fill the transporting vehicle, protection of contents from theft, weather and the like. Another factor is shipping time and the availability of insurance on content delivery amongst other factors. The difficulty of arranging shipment due to strict governmental regulations is another issue. Delivery accommodation, for instance, how many other modes need to be employed apart from the primary mode to get products to the final consumer? Seasonal considerations like weather, flight delays in rainy seasons, ECT. Size of the product to be shipped like cars, small electrical components, ECT, and finally, the perish ability of the product been shipped.

      The contract development process is a key aspect of establishing a solid foundation for a multiyear relationship with an outside vendor. This is a key factor in controlling transportation cost and requires a significant amount of work, diligence and a detailed review of what your transportation needs are. Transportation contributes the highest cost among the related elements in logistics systems, the improvement of negotiations among customers and suppliers should are always evolving; there is always room for improvement. Transportation efficiency could change the overall performance of a company and plays an important role in the overall logistics system, as well as its activities in various sections of the logistics processes. Without the linking of transportation, a powerful logistics strategy cannot bring its capacity into full play, which is where negotiating is a key to success. The transport system provides a clearer notion on transport applications in logistics activities. The development of logistics will be still vigorous in the following decades and the logistics concepts might be applied in more fields. Understanding the entire impact of freight negotiation not only from a shipper's perspective, but also from a carrier's needs can provide a more balanced negotiation and ultimately agreements.

Mathematical Calculations & Answers

How to calculate a weight/measure rate

Commonly used formulas

L x W x H x # of pieces divided by 1728 = Cubic Feet

Cubic Feet divided by 35.314 = Cubic Meters

Pieces x Weight/Piece = Weight in Pounds

Weight in Pounds divided by 2.2046 = Weight in Kilos

Example 1

-Nine pallets, each 150kgs and 122cm x 101.5cm x 127cm (English Standard Measure, each

330.7lbs and 48in x 40in x 50in)

- 9 pallets x 122cm x 101.5cm x 127cm / 1,000,000 cubic centimeters = 14.15 cubic meters

Or

9 pallets x 48in x 40in x 50in = cubic inches / 1,728 = cubic feet / 35.314 = 14.15 cubic

meters

The physical weight of this shipment is 9 pallets x 150 kilos = 1,350 physical kilos. For the volume of this cargo not to exceed the physical weight, the physical weight would need to be at least 14,150 kilos.

Since this is not the case, the ocean freight would be calculated based on 14.15 cubic meters.

Example 2

- Let's imagine a shipment of 10 boxes of a uniform shape of 40" x 40" x 40" and a uniform weight of 200 pounds each. The W/M rate is $80.00. Now, let's calculate the actual charge for this shipment.

- When you see W/M, it means weight or measure (whichever is greater). For LCL Ocean,

Weight is 1000KG increments and Measure is CBM.

- In our example, we would have 10 x 40 x 40 x 40 divided by 1728 = 370.37 Cubic Feet

370.37 divided by 35.314 = 10.49 CBM

- We have 200lbs x 10 = 2000lbs 2000 divided by 2.2046 = 907.19 KGS

This shipment has 10.49 Measure Units and .907 Weight Units and is much "fluffier" than it is

heavy. The ocean freight would be calculated based on 10.49 cubit meters. At

$80 W/M, the ocean freight will be $839.20.

Holt-Winters' Exponential Smoothing Example

Forecast Summary Section

Variable   Pulp Price

Number of Rows   84

Mean   579.2857

Pseudo R-Squared   0.766036

Mean Square Error   4904.916

Mean |Error| 44.74108

Mean |Percent Error| 7.992905

Forecast Method   Winter's with multiplicative seasonal adjustment.

Search Iterations   120

Search CriterionMean |Percent Error|

Alpha   0.999787

Beta   0.1984507

Gamma   0.4674903

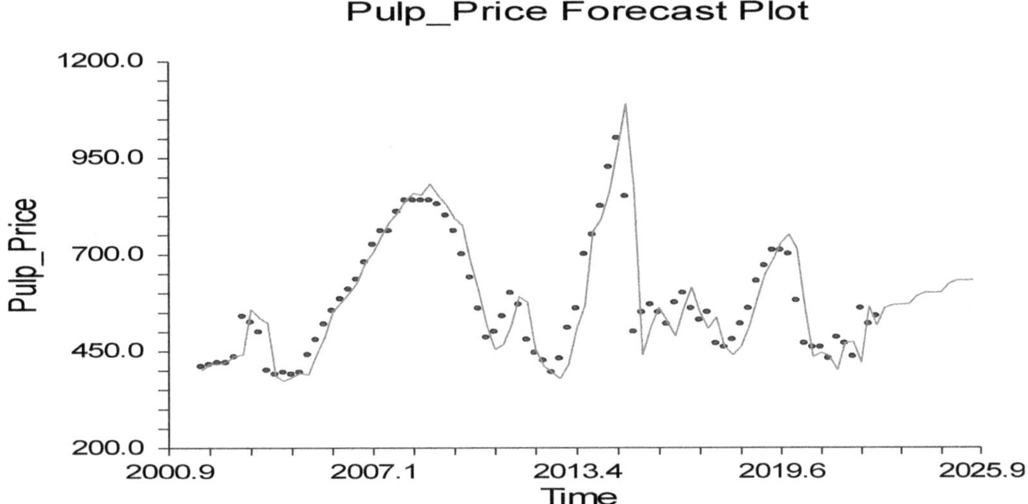

Initial values for forecasts

Intercept (A)   -113.6628

Slope (B)   7.878917

Season 1 Factor 1.008922

Season 2 Factor 0.9970459

Season 3 Factor 0.9850978

Season 4 Factor 1.008935

Review Questions/Answer

1. What are the current available modes of transportation?

Air

Pipeline

Rail

Road

Water

2. What are the inherent costs that are associated with the various modes of transportation? Describe each.

Terminal costs- These are the costs associated with the loading and unloading of product. They are essentially unavoidable.

Line haul costs- These are the costs associated with the function of hauling x amount of freight over a certain amount of distance (fuel, labor).

Capital costs- Costs related to the physical components of the transportation (the actual vehicles, terminals and infrastructure along the way). Physical assets are not forever and can deteriorate. Capital costs are necessary over time, but are not an every trip type of cost.

3. Describe "Intermodal" transportation.

The term intermodal has been used in many applications that include passenger transportation and the containerization of freight. A more descriptive term for this process would be multimodal, because of a lack of effective and efficient connectivity for both freight and information among and between the various modes on shipments under a single freight bill, intermodal freight transport is defined as the use of two or more modes to move a shipment from origin to destination. An intermodal movement involves the physical infrastructure, goods movement and transfer, and information drivers and capabilities under a single freight bill encompassing all single-bill shipments using multiple modes.

4. Why is using an RFP a more effective option in the bid process?

The more effective of the two options is to use a bid process by submitting an RFP. By using the bid process, logisticians may be able to determine what kind of price the market will yield for the transport of their product by receiving so many returns. Because there is no limit to how many RFPs that can be sent, a greater understanding can be achieved. Since the market for transportation services is so competitive, the bid process can cause competitors to try and seize their opportunities for the business by submitting the lowest possible prices. %.

5. What are some of the misconceptions involved with rate negotiation, explain?

A better rate isn't necessarily a good rate.

 A firm can't know if they are getting competitive freight rates unless they know what other shippers are paying.

To carriers, all freight is not the same.

 Knowing what freight the carriers want can earn rate reductions. Sales incentives change regularly, so understanding type of freight certain carriers seek is critical. At times splitting freight across multiple carriers to gain the very best rate and service is to a firm's advantage.

Bundling all your freight with one carrier doesn't ensure the best pricing.

 This is a common misconception with consultants that negotiate freight. Because they think negotiating freight is like negotiating pencils, they miss the nuances of the commodity and leave money on the table.

You need high volume and leverage to get the best rates.

        Many third party logistics companies and brokers will tell a shipper that it is through   leverage that they are able to get better rates for their customers. Leverage doesn't drive down rates, but a deep and extensive knowledge of the industry, transportation   management, and knowing what can be negotiated, will. Having extensive knowledge,   expertise, and experience drives better rates.

Third Party Logistics Companies or brokers will save you money versus what you can do yourself.

        Consider how most brokers and 3PLs make their money. They mark up the difference   between the rate they negotiate with the carrier and how much they charge. Their goal is   to charge as much as possible while creating a perception that they are saving money.   This explains why when one broker is competing for business with another broker the   saving spread is so minimal.

        You will always get the best rates when going directly to the carriers.

If a business opts to work with and negotiate with carriers directly, they need to have extensive industry knowledge as we said before. They need to understand base rate pricing, surcharges, and what the carrier is willing to negotiate. Carriers are experts at creating the perception that a shipper is getting the best rates possible simply by elevating their base rate and then offering deep discounts. Having deep and up-to-date knowledge of the industry can help a business save money when negotiating directly with carriers, however, the cost of handling logistics in-house can be expensive.

References

1. Ballou, R. (2004). Business Logistics/Supply Chain Management, 5th Ed. Upper Saddle River, New Jersey: Pearson Education, Inc.

2. Dickin, Peter, et. al. (1990). Location in Space, 3rd Ed. New York: Harper Collins Publishers

3. Freight Transportation. Retrieved from http://www.c2es.org/technology/factsheet/FreightTransportation

4. Notteboom, T. & Rodrigue J.P. Transport Costs. Retrieved from http://people.hofstra.edu/geotrans/eng/ch7en/conc7en/ch7c3en.html

5. Rattet, S. How to Negotiate the Best Transportation Pricing. Retrieved from http://www.associated-online.com/rfpwilliams.htm

6. P. Kotler, Marketing Management: An Asian Perspective, Singapore: Prentice-Hall Inc., 2000.

7.]D. B. Clark, "Is CRM in Your Company's Future," Trusts and Estates, Vol.139, No.6, 2000, pp. 20-24.

8. La Londe, B.J. and M.C. Cooper. Partnerships in Providing Customer Service: A Third Party Perspective, The Council of Logistics Management, Cincinatti, OH, 1989.

9. Abdelwahab, W. M., & Sargious, M. (1990). Freight rate structure and optimal shipment size in freight transportation. Logistics and Transportation Review, 26(3).

10. Locklin, D. P. (1966). Economics of transportation (No. Textbook).

11. Stank, T. P., & Goldsby, T. J. (2000). A framework for transportation decision making in an integrated supply chain. Supply Chain Management: An International Journal, 5(2), 71-78.

LECTURE 11

POWER POINT HANDOUTS

Transport Decisions

Transport Decisions
in Transport Strategy

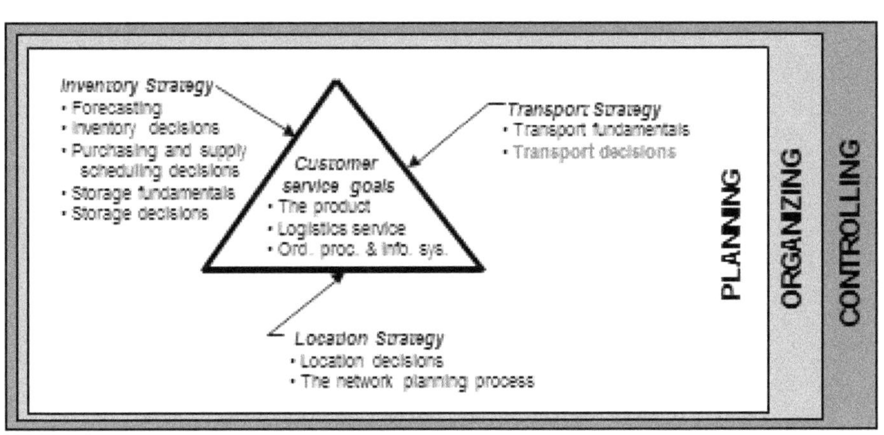

269

Typical Transport Decisions

- *Mode/Service selection*

- *Private fleet planning*
    - Carrier routing
    - Routing from multiple points
    - Routing from coincident origin-destination points
    - Vehicle routing and scheduling

- *Freight consolidation*

Mode/Service Selection

- *The problem*
    - Define the available choices
    - Balance performance effects on inventory against the cost of transport

- *Methods for selection*
    - Indirectly through network configuration
    - Directly through channel simulation
    - Directly through a spreadsheet approach as follows:

```
                                    Alternatives
Cost types                      Air   Truck  Rail
Transportation
In-transit inventory
Source inventory
Destination inventory
```

Mode/Service Selection (Cont'd)

Example Finished goods are to be shipped from a plant inventory to a warehouse inventory some distance away. The expected volume to be shipped in a year is 1,200,000 lb. The product is worth $25 per lb. and the plant and carrying costs are 30% per year.

Other data are:

Transport choice	Rate, $/lb.	Transit time, days	Shipment size, lb.
Rail	0.11	25	100,000
Truck	0.20	13	40,000
Air	0.88	1	16,000

Transport Selection Analysis

Cost type	Compu-tation	Rail	Truck	Air
Trans-portation	RD	.11(1,200,000) = $132,000	.20(1,200,000) = $240,000	.88(1,200,000) = $1,056,000
In-transit inventory	$\dfrac{ICDT}{365}$	[.30(25) ×1,200,000(25)]/365 = $616,438	[.30(25)× 1,200,000(13)]/365 = $320,548	[.30(25)× 1,200,000(1)]/365 = $24,658
Plant inventory	$\dfrac{ICQ}{2}$	[.30(25)× 100,000]/2 = $375,000	[.30(25)× 40,000]/2 = $150,000	[.30(25)× 16,000]/2 = $60,000
Whse inventory	$\dfrac{IC'Q}{2}$	[.30(25.11)× 100,000]/2 = $376,650	[.30(25.20)× 40,000]/2 = $151,200	[.30(25.88)× 16,000]/2 = $62,112
Totals		$1,500,088	$ 861,748	$1,706,770

Improved service →

Carrier Routing

- Determine the best path between origin and destination points over a network of routes

- Shortest route method is efficient for finding the minimal cost route

- Consider a time network between Amarillo and Fort Worth. Find the minimum travel time.

- The procedure can be paraphrased as:
  - Find the closest unsolved node to a solved node
  - Calculate the cost to the unsolved node by adding the accumulated cost to the solved node to the cost from the solved node to the unsolved node.
  - Select the unsolved node with the minimum time as the new solved node. Identify the link.
  - When the destination node is solved, the computations stop. The solution is found by backtracking through the connections made.

Carrier Routing (Cont'd)

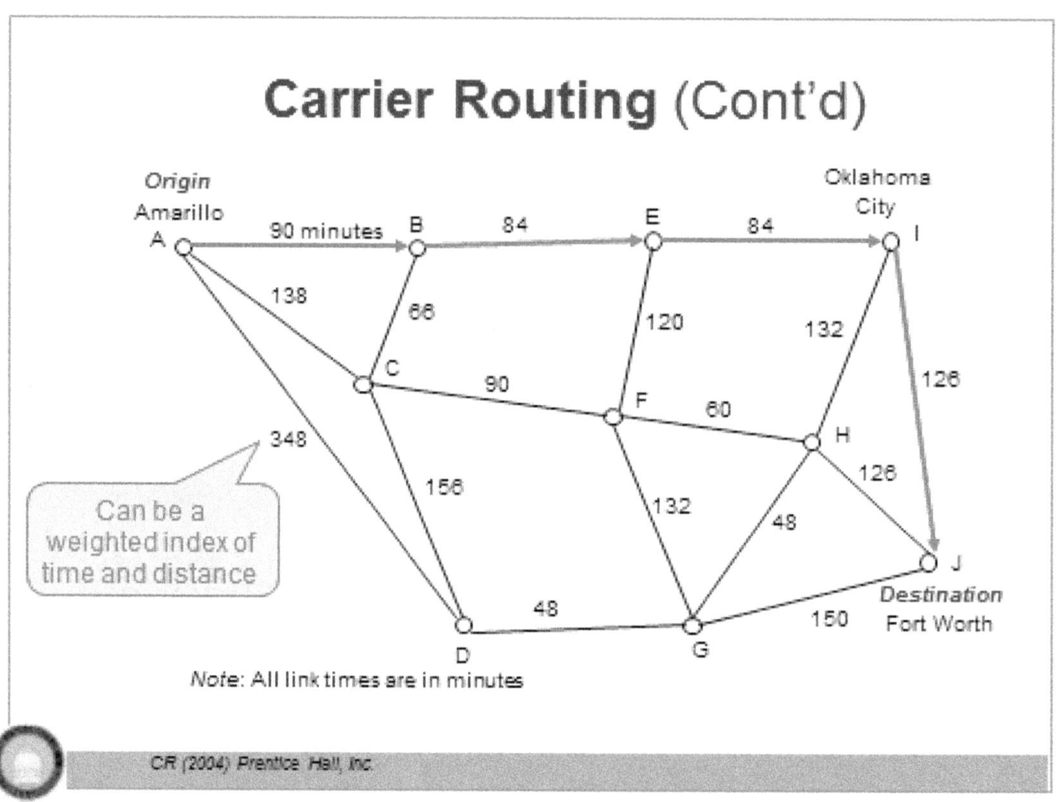

Note: All link times are in minutes

Step	Solved Nodes Directly Connected to Unsolved Nodes	Its Closest Connected Unsolved Node	Total Cost Involved	nth Nearest Node	Its Minimum Cost	Its Last Connection
1	A	B	90	B	90	AB*
2	A	C	138	C	138	AC
	B	C	90+66=156			
3	A	D	348			
	B	E	90+84=174	E	174	BE*
	C	F	138+90=228			
4	A	D	348			
	C	F	138+90=228	F	228	CF
	E	I	174+84=258			
5	A	D	348			
	C	D	138+156=294			
	E	I	174+84=258	I	258	EI*
	F	H	228+60=288			
6	A	D	348			
	C	D	138+156=294			
	F	H	228+60=288	H	288	FH
	I	J	258+126=384			
7	A	D	348			
	C	D	138+156=294	D	294	CD
	F	G	288+132=360			
	H	G	288+48=336			
	I	J	258+126=384			
8	H	J	288+126=414			
	I	J	258+126=384	J	384	IJ*

MAPQUEST SOLUTION

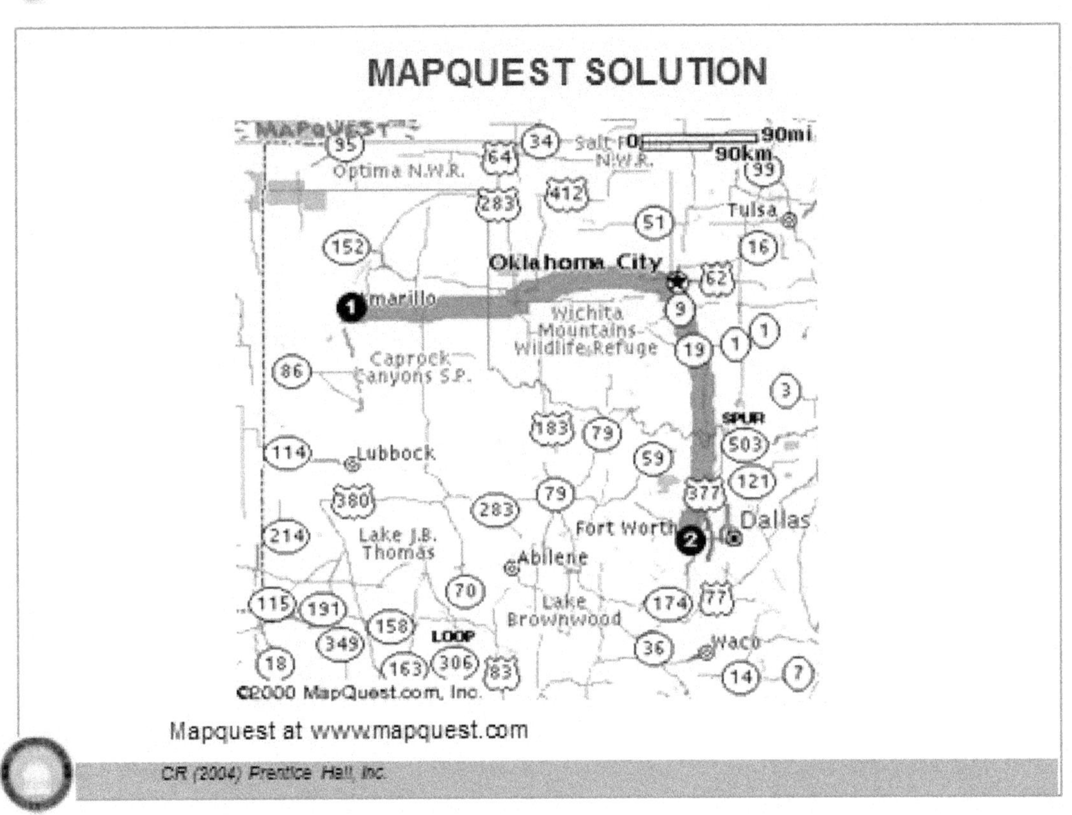

Mapquest at www.mapquest.com

Routing from Multiple Points

This problem is solved by the traditional transportation method of linear programming

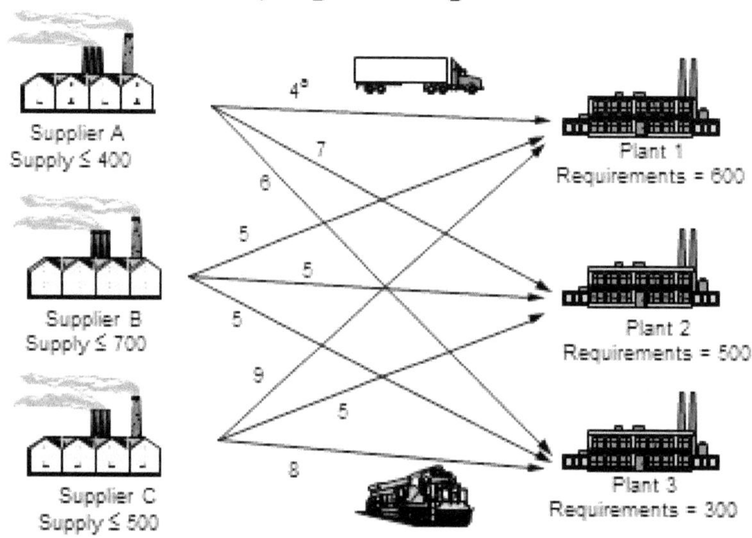

*The transportation rate in $ per ton for an optimal routing between supplier A and plant 1

TRANLP problem setup

From\To	Plt 1	Plt 2	Plt 3	Supply
Sup A	4	7	6	400
Sup B	5	5	5	700
Sup C	9	5	8	500
Demand	600	500	300	

Solution

Results Total cost: 6600

From\To	Plt 1	Plt 2	Plt 3	Supply
Sup A	400	0	0	400
Sup B	200	200	300	700
Sup C	0	300	0	500
Demand	600	500	300	

Routing with a Coincident Origin/Destination Point

- Typical of many single truck routing problems from a single depot.

- Mathematically, a complex problem to solve efficiently. However, good routes can be found by forming a route pattern where the paths do not cross — a "tear drop" pattern.

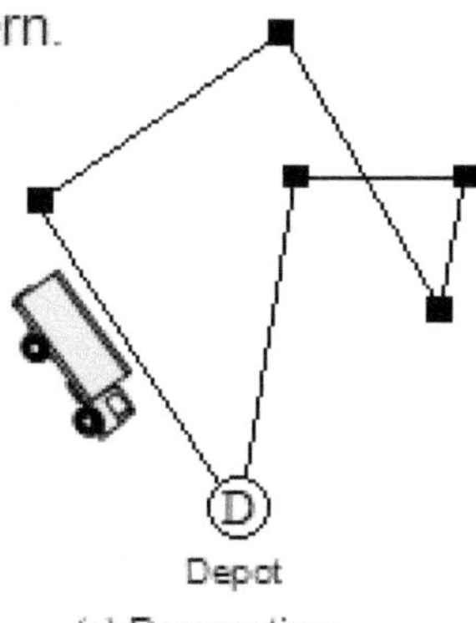

(a) Poor routing--
paths cross

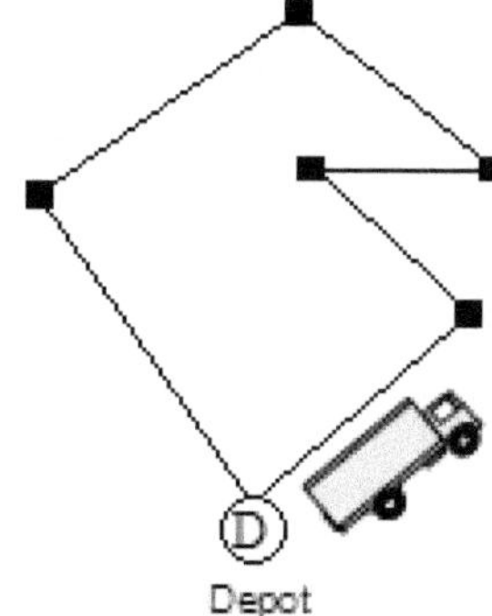

(b) Good routing--
no paths cross

Single Route Developed by
ROUTESEQ in LOGWARE

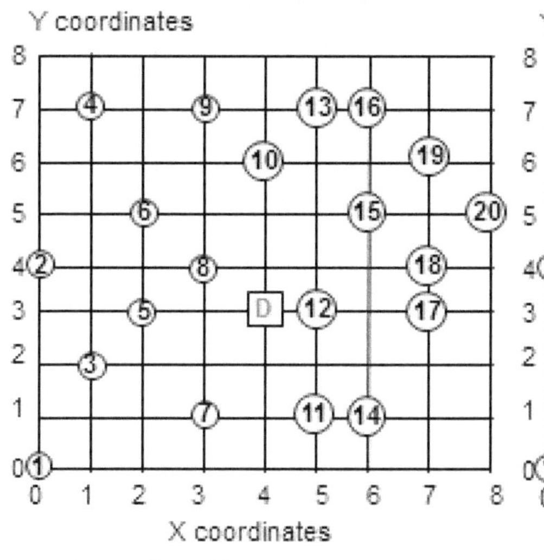

(a) Location of beverage accounts and distribution center (D) with grid overlay

(b) Suggested routing pattern

Multi-Vehicle Routing and Scheduling

- A problem similar to the single-vehicle routing problem except that a number of restrictions are placed on the problem. Chief among these are:

    - A mixture of vehicles with different capacities
    - Time windows on the stops
    - Pickups combined with deliveries
    - Total travel time for a vehicle

Practical Guidelines for Good Routing and Scheduling

1. *Load trucks with stop volumes that are in closest proximity to each other*

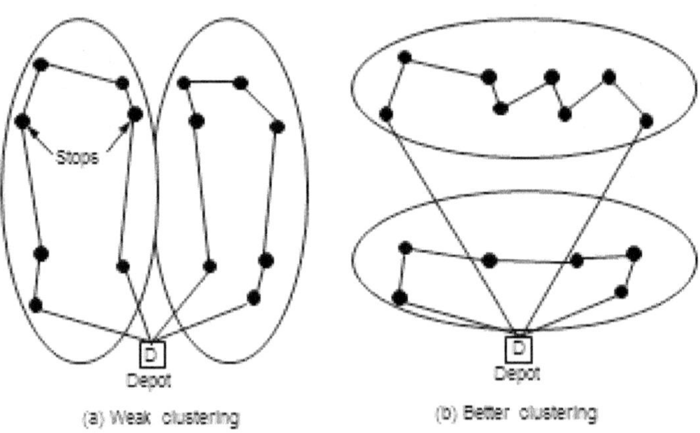

Guidelines (Cont'd)

2. *Stops on different days should be arranged to produce tight clusters*

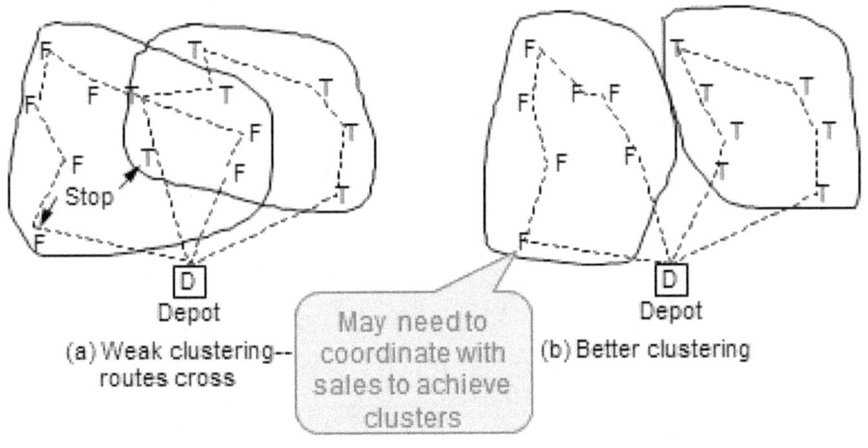

Guidelines (Cont'd)

3. *Build routes beginning with the farthest stop from the depot*

4. *The stop sequence on a route should form a teardrop pattern (without time windows)*

5. *The most efficient routes are built using the largest vehicles available first*

6. *Pickups should be mixed into delivery routes rather than assigned to the end of the routes*

7. *A stop that is greatly removed from a route cluster is a good candidate for an alternate means of delivery*

8. *Narrow stop time window restrictions should be avoided (relaxed)*

Application of Guidelines to Casket Distribution

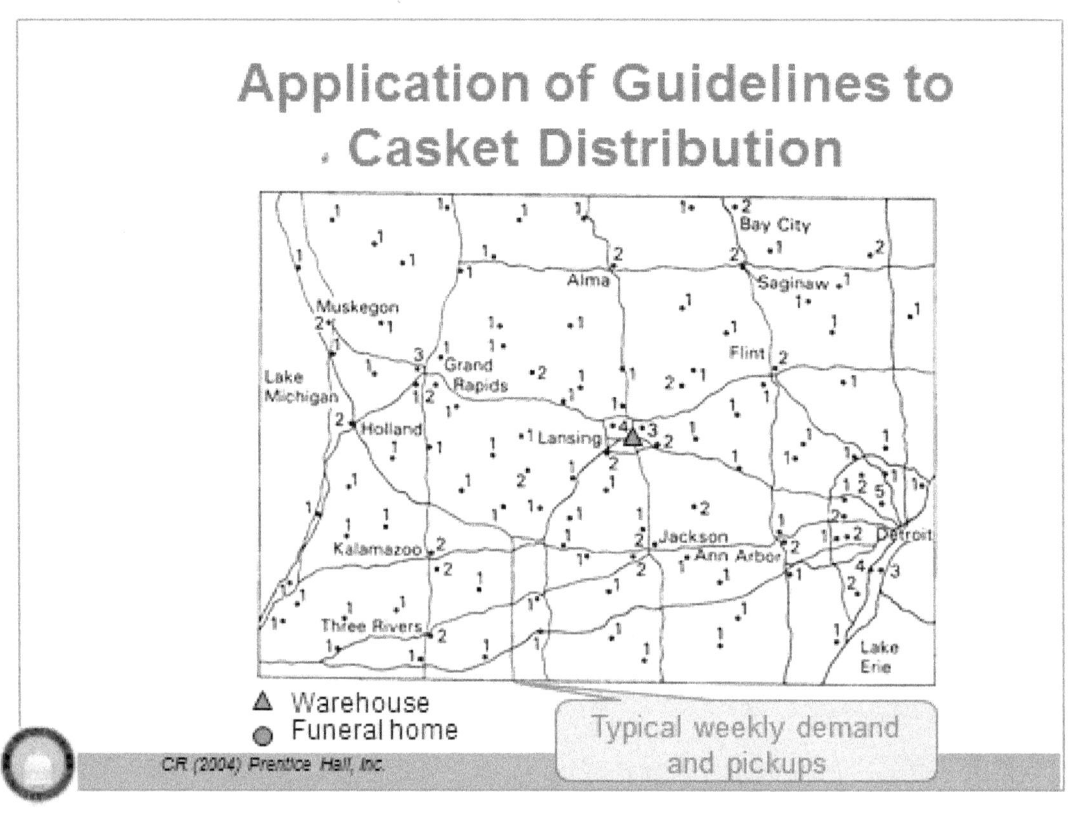

CR (2004) Prentice Hall, Inc.

Application of Guidelines to Casket Distribution (Cont'd)

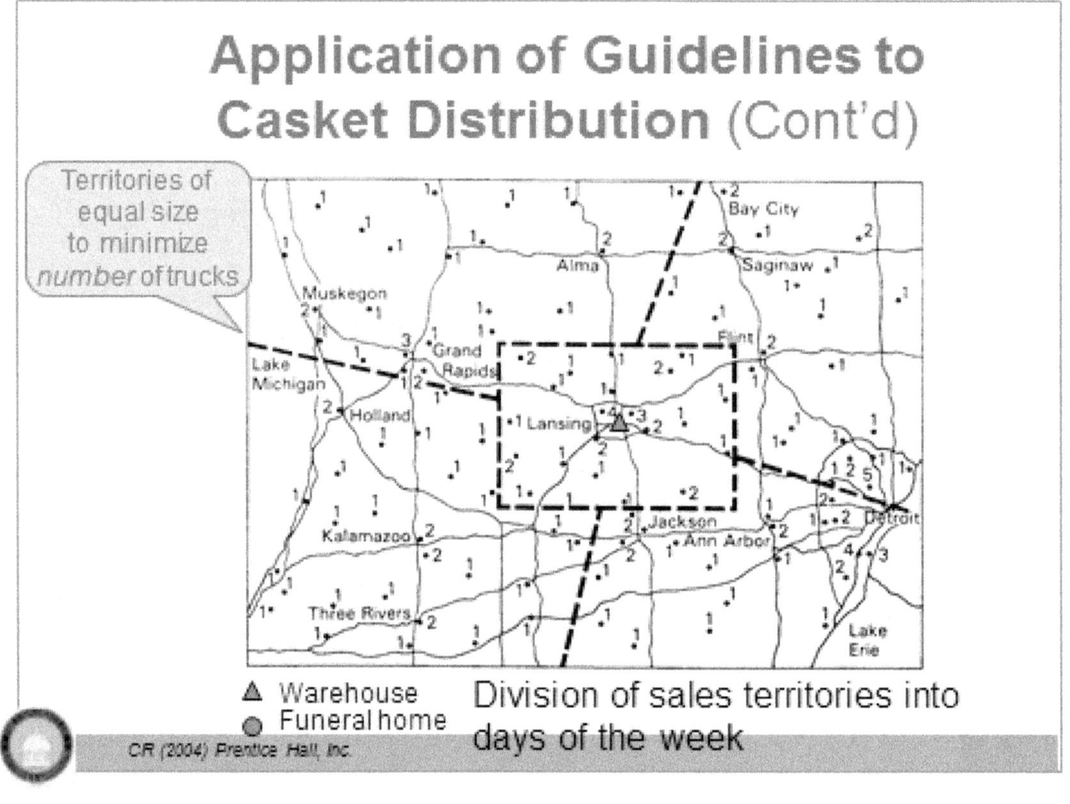

CR (2004) Prentice Hall, Inc.

Application of Guidelines to Casket Distribution (Cont'd)

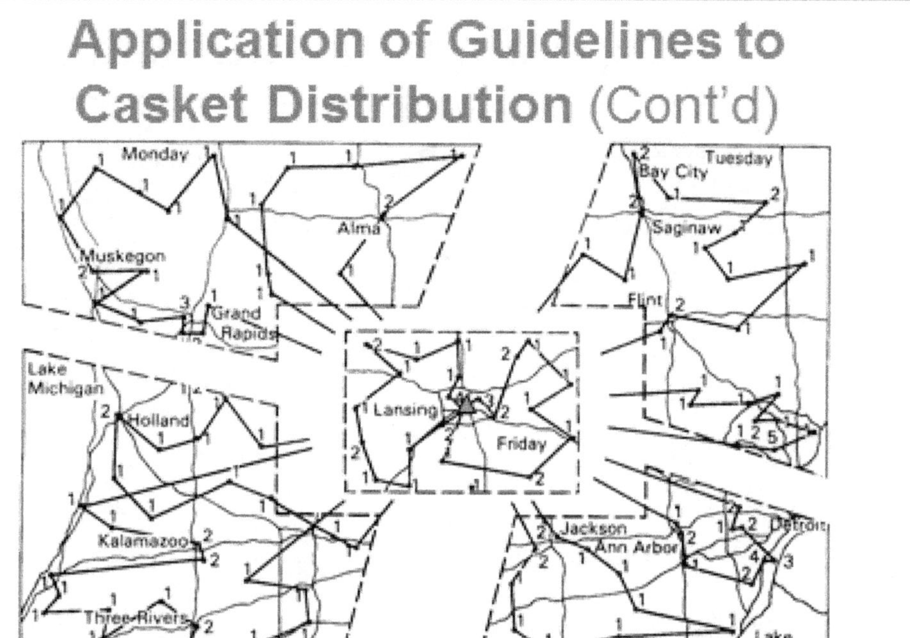

Route design within territories

"Sweep" Method for VRP

Example A trucking company has 10,000-unit vans for merchandise pickup to be consolidated into larger loads for moving over long distances. A day's pickups are shown in the figure below. How should the routes be designed for minimal total travel distance?

Stop Volume and Location

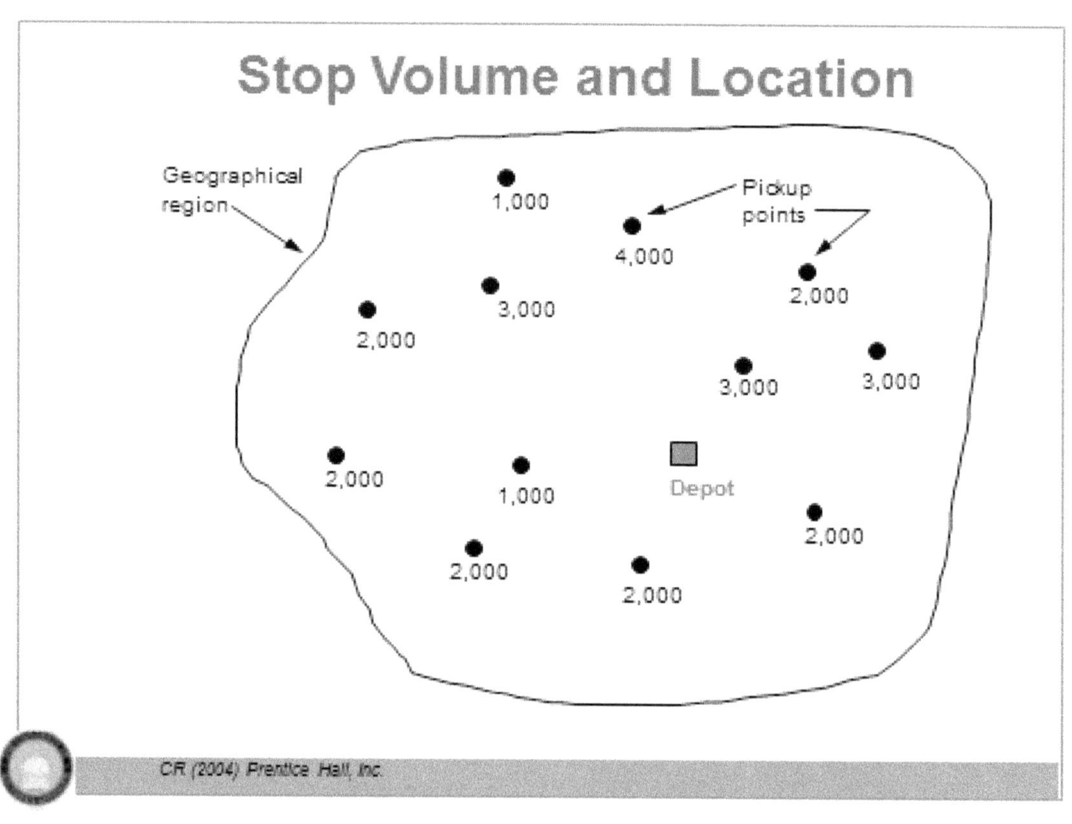

"Sweep" Method Solution

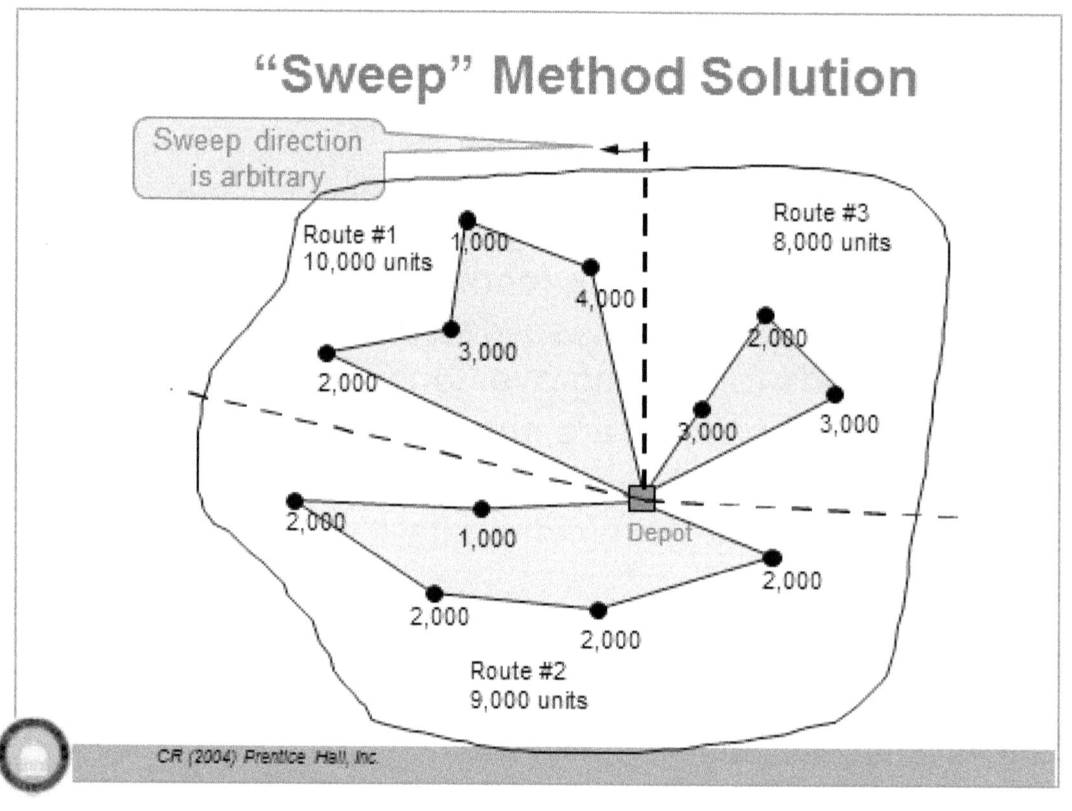

The "Savings" Method for VRP

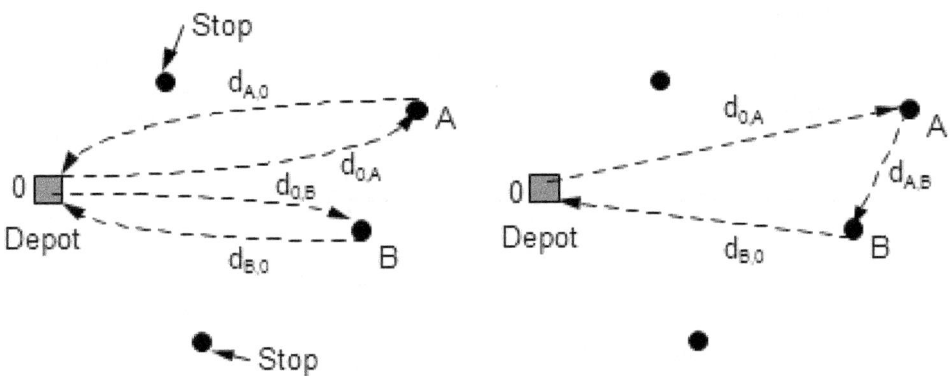

(a) Initial routing—
Route distance = $d_{0,A} + d_{A,0} + d_{0,B} + d_{B,0}$

(b) Combining two stops on a route—
Route distance = $d_{0,A} + d_{A,B} + d_{B,0}$

"Savings" is better than "Sweep" method—has lower average error

Savings Method Observation

The points that offer the greatest savings when combined on the same route are those that are farthest from the depot and that are closest to each other.

This is a good principle for constructing multiple-stop routes

Route Sequencing in VRP

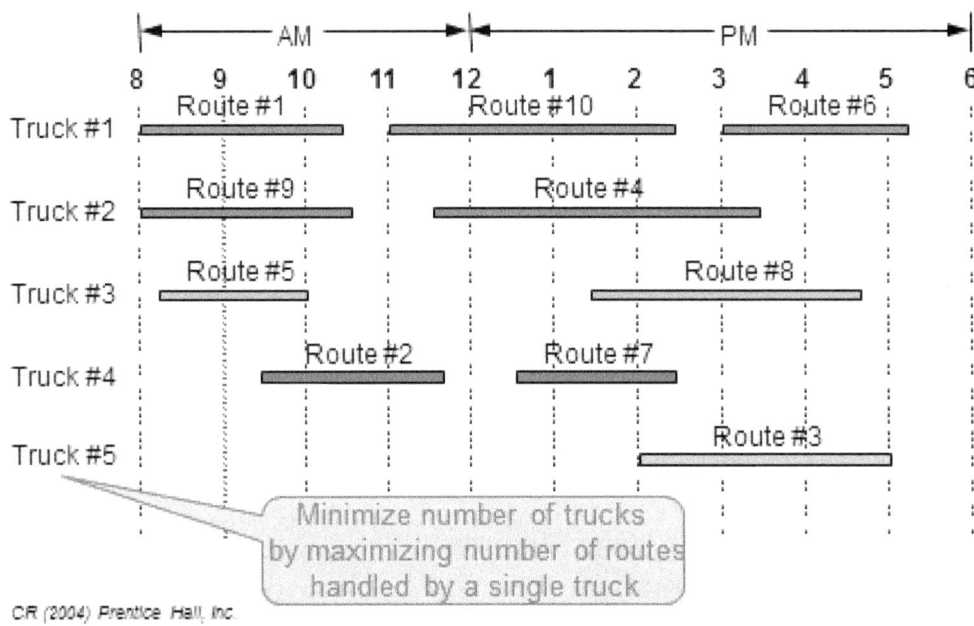

Freight Consolidation

- Combine small shipments into larger ones

- A problem of balancing cost savings against customer service reductions

- An important area for cost reduction in many firms

- Based on the rate-shipment size relationship for for-hire carriers

Freight Consolidation Analysis

Suppose we have the following orders for the next three days.

From:

	Day 1	Day 2	Day 3
Ft Worth			
To: Topeka	5,000 lb.	25,000 lb.	18,000 lb.
Kansas City	7,000	12,000	21,000
Wichita	42,000	38,000	61,000

Consider shipping these orders each day or consolidating them into one shipment. Suppose that we know the transport rates.

Note: Rates from an interstate tariff

Freight Consolidation Analysis (Cont'd)

Separate shipments

	Day 1 Rate x volume = cost	Day 2 Rate x volume = cost
Topeka	3.42 x 50 = $171.00	1.14 x 250 = $285.00
Kansas City	3.60 x 70 = 252.00	1.44 x 120 = 172.80
Wichita	0.68 x 420 = 285.60	0.68 x 400[a] = 272.00
	Total $708.60	Total $729.80

[a] Ship 380 cwt., as if full truckload of 400 cwt.

	Day 3 Rate x volume = cost	Totals
Topeka	1.36 x 180 = $244.80	$700.80
Kansas City	1.20 x 210 = 252.00	676.80
Wichita	0.68 x 610 = 414.80	972.40
	Total $911.60	$2,350.00

Freight Consolidation Analysis (Cont'd)

Consolidated shipment

Computing transport cost for one combined, three-day shipment

	Day 3
	Rate x volume = cost
Topeka	0.82 x 480[a] = \$393.60
Kansas City	0.86 x 400 = 344.00
Wichita	0.68 x 1410 = 958.80
	Total \$1,696.40

[a] 480 = 50 + 250 + 180

Lecture 11
Reading Materials

11.0 Introduction Transportation Services

Supply *chains cannot tolerate even 24 hours of disruption. So if you lose your place in the supply chain because of wild behavior you could lose a lot. It would be like pouring cement down one of your oil wells.*
– Thomas Friedman

In today's global market, the logistics of moving goods around the world is the most fundamental aspect of the local economy. Finding the right transportation mode is more important than ever because transportation costs absorb between one-third and two-thirds of total logistics costs. Logisticians need to understand the role of each transportation mode, compare and contrast each mode and find the most efficient method of transport at the least possible cost. Effective supply chain management depends on understanding how to choose the right form of transport for whatever freight you are moving. Transport fundamentals are the key to successful supply chain modeling. The following chapter will discuss transportation decisions including service choices, transport costs, rates, and transportations documentation such as Bill of Lading and international transport documents. Each section will explain these decisions accompanied by graphical representations of some of these aspects.

11.1 Definition-

There are many accepted definitions for the word logistics, ranging from highly complex to relatively simple. According to Merriam-Webster, logistics deals with the aspect of military science dealing with the procurement, maintenance, and transportation of military materiel, facilities, and personnel; the handling of the details of an operation. Today, logistics does not exclusively pertain to military personnel. Logistical operations can be seen in many industries, from the procurement of materials to third-party logistics and brokering.

11.2 Importance of Choosing the Most Effective Transportation System

Choosing the correct method of transport is crucial to getting the freight where it needs to go, when it needs to be there. There are many forms of transportations including railway, waterway, air and, the most popular, trucking. Sometimes it's a matter of choice, but other times there is no other alternative. For instance, if we are shipping freight from the United States to Tasmania we only have two choices, waterway or air. Obviously, we cannot truck it or put it on rail to Tasmania until they build trucks and trains that can either float or fly! Logisticians want to find the most effective method at the least possible cost. Efficient movement of goods includes 5 criteria or the 5 "rights"; right place/person, right time, right product, right condition and right price. Much transportation today is considered intermodal transportation. Intermodal transportation by definition is transportation involving more than one mode of transportation such as rail-motor, motor-air or rail-water. Using intermodal transportation can usually help cut some of the transport costs versus using just one method. The following graph will help you better understand transportation options by comparing each mode of transport.

	Truck	Rail	Air	Water
Operational Cost	Moderate	Low	High	Low
Market Coverage	Pt-Pt	Terminal-Terminal	Terminal-Terminal	Terminal-Terminal
Degree of Competition	Many	Few	Moderate	Few
Traffic Type	All Types	Low to Mod Value/Mod to High Density	High Value, Low Density	Low Value, High Density
Length of Haul	Short-Long	Medium-Long	Long	Medium-Long
Capacity (tons)	10 to 25	50 to 12,000	5 to 12	1,000 to 6,000
Speed	Moderate	Slow	Fast	Slow
Availability	High	Moderate	Moderate	Low
Consistency	High	Moderate	Moderate	Low
Loss & Damage	Low	High	Low	Moderate
Flexibility	High	Low	Moderate	Low
BTU/Ton-Mile	2,800	670	42,000	680
Cents/ Ton-Mile	7.5	1.4	21.9	0.3
Avg. Length of Haul	300	500	1000	1000
Avg. Speed (MPH)	40	20	400	10

Table I: Transportation Options- Mode Comparisons

II.3 Service Choices

Now let us examine some service choices. Users of transportation have a wide range of services at their disposal which revolve around 5 basic modes: rail, truck, water, air and pipeline. We will discuss pipeline in more detail in a later chapter, so our focus primarily will be the other four basic modes of transport. All of these modes, however, have some commonality to each other. Five basic characteristics are common in all transport modes and those are price, average transit time, transit time variability, and loss and damage. These factors are the most important aspects to all transport decision makers. Let's look at each one individually.

II.3.I Characteristics

Price or the cost of the transport service to the shipper is simply called the line-haul rate. In my experience in the trucking business, if price is not the most important aspect then it definitely comes in a close second. We always tell our new hires that they must remember three things when sealing the deal and that's weight, rate and date. That is another story though. Price can make or break a deal and transport service costs can vary greatly among different modes. Transit time is also another important characteristic. If a customer wants a product in a short amount of time, he may choose to pay for air transport, versus, say, for instance, water transport. Finally, another important characteristic is loss and damage. This can be very important when making carrier selection. Delayed shipments and goods arriving in unusable condition can greatly inconvenience the customer. This could cause higher inventory costs, stock outs, a greater number of backorders, etc. Choosing the right form of transport is critical. For example, a train may not be the best way to ship breakable goods like fine china, but is the perfect way to transport coal and raw materials. On the flip side, transporting raw materials by plane does not make any sense because they are so heavy and there is no need to transport raw materials over long distances by air. The next few sections well discuss some different service choices, from single-service to intermodal and everything in between.

11.3.2 Single Service Choices

Each transportation mode offers its services directly to the user. We call these choices *single service* because there is no intermodal services provided; once a product gets on train, plane, truck or ship it stays their until it reaches its final destination. Single service choices usually cut out the "middle men" such as brokers and freight forwarders by offering services directly to the customer. Let's examine each of the four main forms of freight transport and what single service choices in freight are commonly transported on them.

<u>Rail</u> – long hauler and slow moving; usually travels at 20 MPH and average haul is 712 miles; usually hauls raw materials such as coal, chemicals, lumber, pulp and other products that usually need further refining.

<u>Truck</u> – moderate moving; can travel up to 680-700 miles per day; usually haul semi-finished and finished goods; most popular form of transport; can haul 40,000 to 48,500 lbs. legally; according to truckinfo.net, the trucking industry employs 8.9 million people in the United States alone, more than 3.5 are truck drivers (others are support staff, mechanics, brokers, etc.).

<u>Air</u> – fast moving; average travel for air freight is more than 1,000 miles; commercial jets can travel up to 585 MPH; air freight transport costs twice as much as trucking and 16x as much as rail; the least reliable mode due to mechanical breakdowns, weather conditions and variability in services.

<u>Water</u> – slowest form of transport; average speed of 5-15 MPH; inexpensive; can haul massive amounts of goods, upwards of 40,000 tons; average haul is up to 500 miles in inner waterways and over 1,000 along coast lines and open water.

11.3.3 Intermodal Services

Intermodal services are those that use more than one form of transport. According to Ronald H. Ballou, there are ten intermodal service combinations: (1) rail-truck, (2) rail-water, (3) rail-air, (4) rail-pipeline, (5) truck-air, (6) truck-water, (7) truck-pipeline, (8) water-pipeline, (9) water-air, and (10) air-pipeline.

<u>Trailer on Flatcar</u> – (TOFC) aka "piggyback"; refers to transporting truck trailers on railroad flatcars. Typically done on longer hauls than normal freight trucks take (*See ex. 1*)

Ex 1- Trailer on Flatcar

<u>Containerized Freight</u> – Under TOFC arrangement, entire trailer is transported using a flatcar; also consider Container on Flatcar (COFC) which is hauling only the container minus the wheels. (*See ex. 2*)

Ex. 2 Container on Flatcar

11.4 Transport Cost

All businesses incur some type of cost, in the transportation industry or service these costs could be classified into two classes or groups called variable or fixed costs. These costs are usually classified as such at the sole discretion of the individual based on his/her perspective. There are always advantages of one mode of transportation against the other modes for cost of services. Due to the difference of cost characteristics one may favor a certain mode over the other at any given time or circumstance. Usually the prices are keyed to the characteristics of each type of service.

II.4.I Cost Classes:

Cost can be categorized or group together. There are two main coat classes namely Variable and Fixed costs and Common or Joint costs.

Variable and Fixed Costs-Variable costs are costs that vary with service and volume. While fixed costs don't vary due to service or volume. Examples of variable costs are fuel, labor, equipment maintenance, handling, pickup and delivery. While fixed costs are roadway acquisition, maintenance, terminal facilities, transports equipment etc.

Common or Joint Costs- Sometimes it is very difficult to allocate cost due to the fact that different products with different weight, size, proportion or shape may all be shipped in one line- haul. Thus to allocate cost to the different products becomes a headache. Common or joint costs can sometimes be very painstaking for the company.

II.4.2 Cost Characteristics by Mode:

This is the transportation costs associated with the type or mode of transportation used.

Rail: It has very low variable costs and high fixed costs. The rail costs are usually categorized into train operational costs, track and signaling costs and terminal and station costs. From a regulatory point of view, the allocation of the different costs that are avoidable and those that are common is necessary [compos and cantos, 1999:5] . The scale and scope economies in the rail industry also create problems for regulation. The most notable problem of scale and scope economies is the fact that it is not easy to allocate costs in the rail industry [Kessides and Willig, 1995:7]

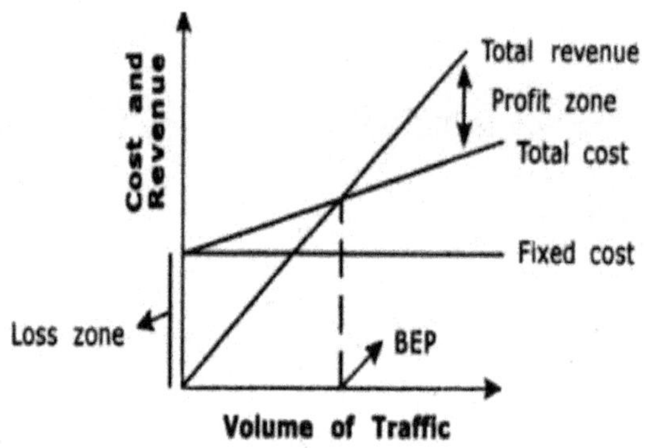

FIGURE 4.9 Railways Cost Characteristics.

Highway: This carries the lowest fixed costs compared to other modes of transportation. The variable costs on the other hand can be high for fuel, tolls and weight mile taxes can be high at times. Terminal expenses and line-haul expenses are the two main categories of trucking costs. The terminal expenses usually include pickup and delivery, handling, billing and collecting. The Line hauls costs which are usually 50-60% of the total costs of the highway. According to Ronald H. Ballou it is not clear that per unit, line-haul costs necessarily decrease with distance or volume. However, total unit trucking costs decrease with shipment size and distance as terminal costs and other fixed expenses are spread over more miles but not as dramatically as rail costs.

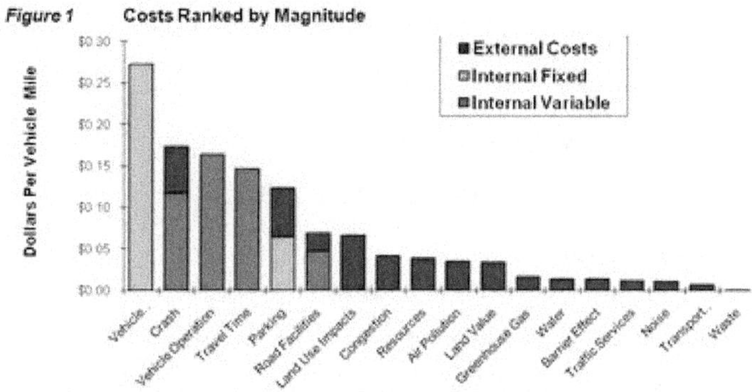

This figure shows costs per vehicle-mile for an average North American automobile.

Water: Most of the fixed costs associated with water carriers are associated with terminal operations. These costs are labor fees and loading and unloading fees. The terminal costs associated with Water transportation are usually high but are offset by the low Line-Haul costs.

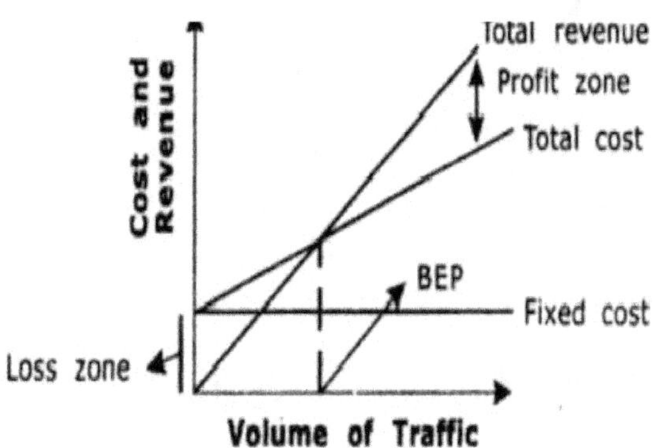

FIGURE 4.10 Waterways Cost Characteristics.

Air: Water and highway transportation have similar cost characteristics as Air transportation. The terminal costs associated with air transportation come in the form of fuel, storage, space rental and landing fees. In the case of freight operations, ground handling and pick-up and delivery can be considered terminal costs. Variable costs associated with Air transportation are reduced by the length of the line haul

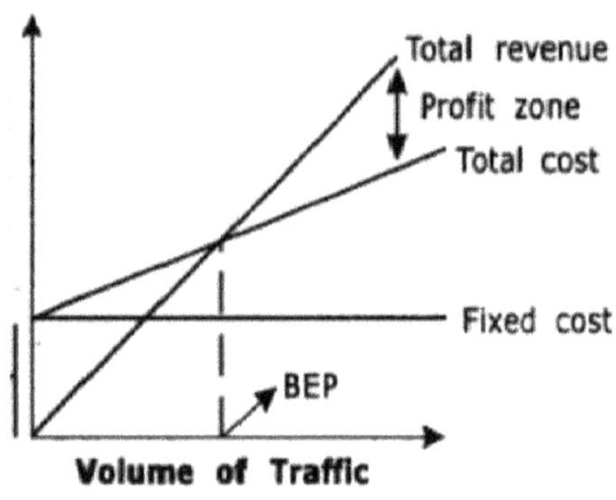

4.11   Airways Cost Characteristics.

Pipeline: It has the highest ratio of fixed costs to total costs of any mode. The high fixed costs such as terminal costs, pumping equipment, are spread through high volume products been transported through the pipeline. The variable costs include the movement of products, operation of pumping stations just to name a few.

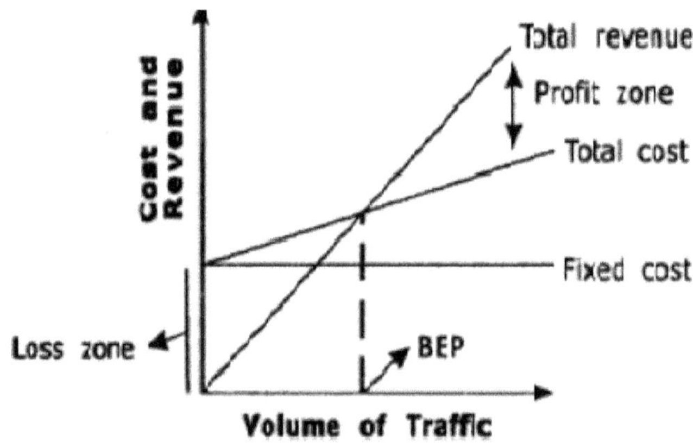

FIGURE 4.12   Pipelines Cost Characteristics.

II.5   Rate Profiles

The prices usually charged by carriers are usually referred to as transportation rates. These rates are usually influenced by the volume, distance and demand and various means are used to calculate the rates. We have volume related rates, distance related rates and Demand related rates.

II.5.I Rate Categories

Volume related rates: These are rates based solely on the size or volume of the shipment. The size of the shipment is directly related most of the time to the cost of the service thus the rate.

Distance related rates: These are rates computed from taking distance as the main factor. Rate structures are usually in the middle of the far left which is directly related to distance and the far right which is indirectly related to distance. It comprises of uniform rates, proportional rates, tapering rates, and blanket rates.

Demand related rates: The rate that is charged can also be dictated by the demand. People view transportation as having a value and are usually prepared to pay amount proportional to that value they attach to it thus rate related to demand adheres to an upper limit.

11.5.2 Line Haul Rates:

These are charges and cost incurred between destination and origin terminals. In the case of motor carrier services it is door to door costs. These costs are usually determined or classified by the size of the shipment, the type of product, by route or miscellaneous.

By Product- A uniform freight classification code was adopted by many railroads, truckers, and water carriers around the world around the 1950s. A variety of factors ids used in determining a product rating. Some of these factors include density ease of handling, stowability, liability etc.

Class Rates- The classification goes hand in hand with tariffs or transportation price lists.

Contract Rates- are also under line haul rates but they generally take precedence over class rates and are usually tailored uniquely to reflect individual shipping situations.

Freight All Kinds- By Shipment Size: rates vary due to shipment size and quantity. They are usually quoted on a dollar per hundred pound (cwt.) bases.

Other Incentive Rates- incentives are given to ship in large quantities. When large quantity shipments are done it warrants incentive rates so as to encourage large shipment sizes.

*Miscellaneous Rates-*are rates that do not fit any particular classification. Some of the rates classified under miscellaneous are

Cube Rates- These are rates based on space occupied rather than weight.

*Import or Export Rates-*These are rates for items or cargo destined for exportation or importation thus they are given special rates in other in other to encourage foreign trade and investment.

Deferred Rates- these are special rates given in other for the carrier to be able to delay shipments to fill out available space. This gives the shipper a lower rate and helps the carrier fill out available space thus been a win-win situation for both parties.

Released value Rates- This rate enables or reduces the liability of the carrier on items or products. The carrier usually has a set amount of liability.

Ocean freight Rates- These are rates which are quoted differently for domestically moved goods. They are usually quoted solely by the discretion of the option of the carrier based on space or weight.

11.5.3 Special Service Charges:

If special services are performed special charges are made. Some of these services are Special Line-Haul services, protection services, terminal services etc.

Special Line Haul services: may include diversion and consignment which is usually changing the destination of a shipment while on route and changing the consignee of the shipment respectively.

Diversion and Consignment – In practical terms diversion and consignment can be the same and they can be used inter changeably. Diversion of a shipment refers to the change of the destination of the shipment when the shipment is already on course. While consignment refers to the change of the individual receiving the shipment or in other words the consignee. Thus to facilitate this process special charges or rates usually apply.

Transit Privileges- In order to prevent the shipper from paying the through rate from the origin of the shipment to the stop off point then to the final destination, transit privileges are accorded to the shipper. They are usually minimal charges for the shipment been stored at a transit point before finally moving it to the final destination.

Protection- Depending on the type of article or goods been shipped some do require some form of protection. Goods such as fruits vegetables or in general perishable goods may require special form of accommodation by the carriers. Thus this usually warrants some type of charges to be added to the overall cost of the shipment.

Interlining- In the event that a certain carrier does not service a particular region, It can contract another carrier that does to move a particular load for them. The first carrier usually pays the second carrier but the customer pays the first carrier the total cost which usually includes the profit of both carriers. This is usually very costly compared to if there was a direct service carrier to that region.

Terminal Services- Charges may apply for services rendered in and around the terminal. Most of these usually fall under pickup and delivery, switching and Demurrage and detention.

Pickup and Delivery- Special charges may be charged for the pickup and delivery of shipment from or to the terminal.

Switching- This is very similar to pickup and delivery but it only applies to rail terminals.

Demurrage and Detention- This are penalties levied on the customer for holding or using equipment or facilities beyond the allocated free time.

11.6 Documentation

There are three different types of documentation types in domestic freight transportation. These include bill of lading, freight bill, and the freight claim. International transportation usually use bill of lading freight bill way bill freight claim etc.

Bill of Lading

The bill of lading usually called by its abbreviations B.O.L is the main document used to facilitate the movement of the freight. It is usually the contract between the shipper and carrier for the safe transportation of freight or goods from origin to destination with a certain reasonable time frame. The B.O.L usually serves three purposes.

I It is usually the receipt for the freight been transported and usually contains all information about the condition of the goods been transported and at what date and time it was or will be transported.

2. It is the binding contract between the shipper and carrier and completely outlines the terms and conditions of the contract.

3. It services as title documentation in the case of a negotiable bill of lading.

11.6.1 Freight Bill

It is an invoice rendered by a carrier to a consignee of freight usually containing a thorough description of the freight, the shipper's name, point where the shipment was shipped from in other words the origin, its weight, and the charges associated with the shipment. It is more thorough than the BOL because the BOL does not usually include the freight charges. Unless credit is available to the shipper the freight charges are usually prepaid by the shipper. Terms of credit usually vary from 48hrs to 7 days depending on the carrier.

11.6.2 Freight Claims

Either due to legal responsibilities as a common carrier or due to overcharges, freight claims are made against carriers.

Loss Damage and Delay Claims-Usually the bill of lading usually stipulates the limits of carrier responsibility if when damage occurs for it's the responsibility of a common carrier to move freight at a reasonable time frame without loss or damage. Loss incurred due to unnecessary and unreasonable delay or failure to meet guaranteed schedules is usually recoverable to the extent of the value reduction resulting from the delay.

Overcharges-Due to incorrect invoices, mathematical errors, duplicate billing, wrong weights, incorrect interpretation of tariff rates etc. can warrant a claim for overcharges from the Shipper to the carrier. Usually a time frame of up to three years is allowed for overcharge claims on interstate shipments.

11.6.3 International Transport Documentation:

The most distinguishable feature between international transportation and domestic is the amount of documentation required for imports and exports. Some of the documents needed or useful during exportation are Bill of lading Dock receipt Delivery Instructions Export Declaration Letter of credit Consular invoice Commercial invoice Certificate of origin Insurance certificate Transmittal Letter. During importation some of the required documents are Arrival notice Customs entries Carrier's certificate and release order Delivery order Freight Release Special Customs invoice. All these documents make it very easy for importation and exportation process documentation.

References

Ballou, R. (2004). *Business logistics/supply chain management.* (5th ed., pp. 164-218). Upper Saddle River, NJ: Pearson Education, Inc.

Dobbins, J., Macgowan, J., & Lipinski, M. (n.d.). Overview of U.S. freight transportation system. (2007). *Center for Intermodal Transportation Studies,*

Gaulier, G., Mirza, D., Turban, S., & Zignago, S. (n.d.). International transportation costs around the world: a new cif/fob rates dataset. (2008).

 SATISH K. KAPOOR, PURVA KANSAL, BASICS OF DISTRIBUTION MANAGEMENT: LOGISTICS APPROACH A

http://www.truckinfo.net/trucking/stats.htm

http://wikipedia.com

I. INTRODUCTION TO TRANSPORTATION ROUTING

Transportation routing is finding the best way to have goods delivered from one point to another. When analyzing a company's transportation one must understand how and why they are routing their freight the way that they are. There may be considerations that all affect the way that the freight can be routed such as cost, delivery times, multiple destinations, and ready time. Transportation Management Systems are designed to keep track of transportation information and help firms make the best possible decisions regarding the movement of goods in their supply chain. "To reduce transportation costs and improve customer service, finding the best paths that vehicle should follow through a network or roads, rail lines, shipping lanes, or air navigational routes that minimize time or distance is a frequent decision problem" (Ballou 225). When a company is looking at their transportation it is best if they gather all relevant data and make the proper decisions based off of what are their most important needs.

This paper will explore some of these decisions. It will look at how to effectively route freight, equipment and manpower, proper scheduling route sequencing, routing systems, and the problems that exist with a transportation routing analysis. Understanding these principles allows decision makers at firms to best utilize precious transportation resources in the most effective way possible. Since transportation is such a large part of logistics costs making proper decisions in utilizing equipment and manpower cannot be overstated.

II. VEHICLE ROUTING

Separate and single origin destination points are the simplest way to route vehicles. One of the main goals of this method is to create a route that best utilizes the vehicle's capacity. Finding the minimum drive time between two points while still meeting the parameters of an associated move is the purpose of this system. It is finding the most efficient way to get between two points. The firm is trying to figure out the exact route that best utilizes the vehicle. When looking at this type of vehicle routing it is important to take into account both what is the most practical route for both distance and quality. Certain routes may have shorter miles but may take more time to traverse. Picking the correct route that balances both distance and time is important for correctly utilizing a firm's assets.

There are many applications that can assist in determining optimal routes. Programs such as PC Miler and IntelliRoute are examples of software that can help determine the correct routes. "The shortest practical route (a blend of distance and time) is the objective of route design" (Ballou 228). These programs assist in these designs. Additionally, they are extremely important because they take into account many other factors that would otherwise be difficult and time consuming for human planners to carry out. These factors include things such as tolls and road construction. "These expanded capabilities have led to reduced rate disputes, reduced fines, and improved audit efficiencies, which in turn result in improved customer service, delivery, reporting, asset utilization and driver retention" (Ballou 228).

A more complex system of vehicle routing is multiple origin and destination points routing. What this system entails is there are multiple pick up locations that may service several delivery locations within the transportation network. The firm must find the origin point that serves the best destination point. They must also find the best routes within the network for doing so. "This problem commonly occurs when there more than one vendor, plant or warehouse to serve more than one customer for the same product. It is further complicated when the source points are restricted to the amount of total customer demand that can be supplied from each location" (Ballou 230). To solve this, firms commonly use an algorithm known as the transportation method. This method takes into account how much suppliers can produce, transportation routes, and request amounts of product by the customer, to create optimal supply routes. Below is an example. It shows all possible variables including how much each plant can produce, transportation cost, and what each supplier is demanding. The firm uses this information to find the best supply route. The diagram indicates the optimal supply routes that best utilize the resources the firm has. Understanding where to source demand from is critical in control logistics costs.

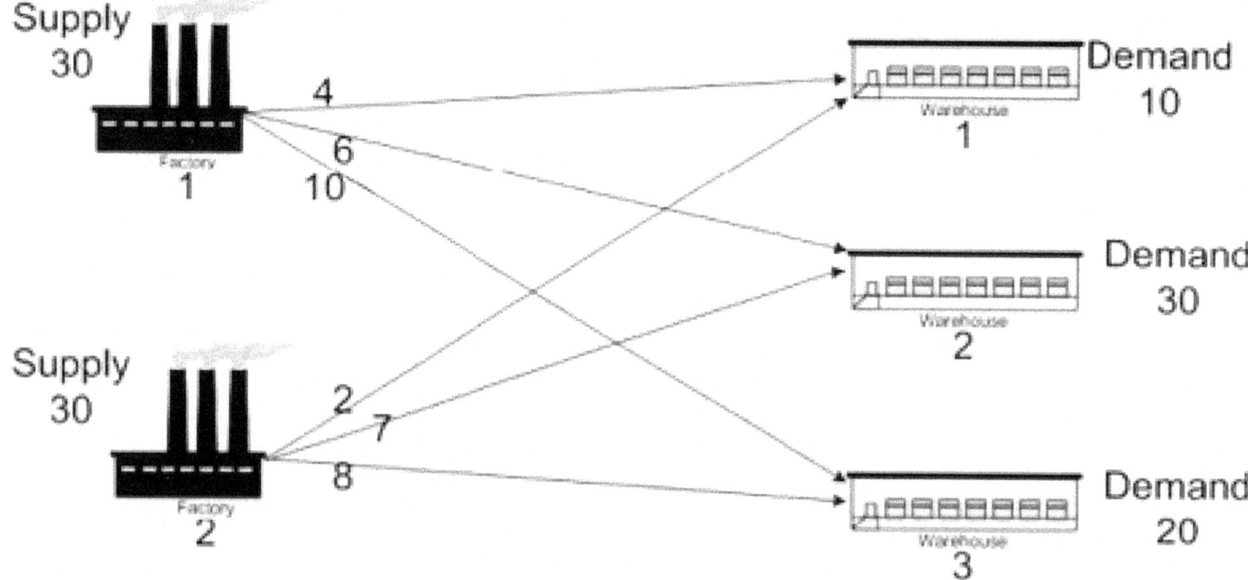

Factory 2 supplies 10 to warehouse I/ Factory 2 supplies 20 to warehouse 3/ Factory I supplies to 30 to warehouse 2

FIGURE I

Being able to understand how to route trucks is an important factor. The individual making decisions must be able to lay out a route that does not cross paths and efficiently gets trucks from one point to another. It is especially important in multiple drop situations that the route takes into account the most effective way to get from point to point because that is the only way a vehicle can be effectively used. Although there may be other factors to consider such as one way streets and poor intersections that having a truck crossing its own path may make sense. Using a computer program that has such knowledge built in to make decisions about when to have a route cross paths is the most effective way to make such a decision.

III. VEHICLE ROUTING AND SCHEDULING

Vehicle routing and scheduling (VRP) takes basic routing principles and expands upon them. It includes real world considerations such as volume, weight restrictions, and drive time limitations of drivers, pickup and delivery considerations, and driver breaks. The optimal solution often does not exist so firms must come up with a best case scenario that meets most of their needs. These solutions require firms to thing logically about their priorities and use the solutions that meet their top needs. There are some very basic principles that a firm can use to decide the best route for trucks. There are seven practices mentioned in *Business Logistics/ Supply Chain Management.* First, are load trucks with stop volumes that are the closest proximity to each other. Stops on different days should be arranged to produce tight clusters. Build routes beginning with the farthest stop from the depot. The sequence of stops on a truck route should form a teardrop pattern. The most efficient routes are built using the largest vehicles available. Pickups should be mixed into delivery routes rather than assigned to the end of routes. A stop that is greatly removed from a route cluster is a good candidate for an alternate means of delivery. Narrow stop time window restrictions should be avoided. All of these principles help a firm maximize its assets.

There are two main methods for routing and scheduling. One is the sweet method and the other is the savings method. The sweep method requires locating all stops on a map. Then one needs to extend straight lines from the depot to all stops in all directions. Understand vehicle capacity and once it is maximized draw another line for another route. Within each route sequence all the stops so that they minimize distance between stops. " The sweep method has the potential of giving very good solutions when each stop volume is a small fraction of the vehicle capacity, all vehicles are the same size, and there are not times restrictions on the routes" (Ballou 243).

The savings method is also a proven and important technique. "The objective of the savings method is to minimize the total distance traveled by all vehicles and to indirectly minimize the number of vehicles needed to serve all stop" (Ballou 243). In this process it is important to continually generate the most condensed route possible so that truck capacity is maximized and efficiencies within the firm are created. These tactics save a firm money.

Route sequencing needs to be put into the practical real world. Sequencing routes make certain that as soon as a truck empties out it can then be utilized again. This means that the truck takes route one at 08.00 then finishes and returns at 10.00. It will then be refilled and go back out at 10.30 on route two and so on. Making sure to meet the needs of a customer is critical. Not all customers take daily or even weekly deliveries. Carefully assigning loads and stops and balancing customer expectations is important so that the size of the truck fleer can be minimized. This is a common problem in areas such as liquor deliveries or food service companies such as Sysco.

IV. THE TRANSPORTATION ROUTING ANALYSIS GEOGRAPHIC INFORMATION SYSTEM

The Transportation Routing Analysis Geographic Information System (TRAGIS) is a potential tool that is used for modeling transportation routing. TRAGIS is a user-friendly, GIS-based transportation and analysis computer model. It offers multiple options for route calculation by utilizing uniquely value added network databases for highway, rail, and waterway infrastructures. It also provides population density data for various transportation segments. The TRAGIS model is deployed as a client-server application, where the map data files and user interface software reside on the user's PC and the routing engine is located on the server. The Arcview software initially formulated TAGRIS, which is versatile multifunctional software. This was upgraded by the C++ programming language. It employs UNIX platform to operate as it incorporates huge routing database.

WEBTAGRIS is the latest user-friendly version, which is accessible for determining routing for the rail, highway, and water transportation modes. It allows the selection of the origin and destination of a route from a list of node names. After an origin and a destination are selected, the model is ready to calculate a route based on criteria established by option selections. A default set of criteria is active for each transportation mode in the model. After completing the route calculation, Web TRAGIS displays the standard route listing. The user can also view a detailed listing of the route and population-density information, which can be used with the RADTRAN risk model. Option settings provide a mechanism to change various parameters used by the model for route calculations. Examples of some of the options include adjusting the penalty factors for the mainline classifications for rail routing, using preferred highway routes for radioactive materials, and running alternative routes for the different transportation modes in Web TRAGIS. It also provides functions to temporarily modify the routing networks. The user can select individual nodes and links or an entire state in which all nodes and links are blocked from the network.
Some of the primary reasons for the development of TRAGIS are (a) to improve the ease of selecting locations for routing, (b) to graphically display the calculated route, and (c) to provide for additional geographic analysis of the route.

TRAGIS features include the ability to:

• Select an origin and destination from a list of city names;
• Automatically calculate alternative transportation routes;
• Modify transportation networks by temporarily blocking nodes, links, railroad companies, or states;
• Calculate highway routes that meet U.S. Department of Transportation regulations for radioactive materials; and identify Indian reservation lands along highway and rail routes

V. FREIGHT CONSOLIDATION

Freight consolidation is the services rendered by shipping companies to reduce the overall shipping cost and increase the shipping security. It is also known as consolidation service, assembly service and cargo consolidation. Due to freight consolidation the companies started shipping their shipments in large quantities. Hence several small shipments are moved to the same location and they are bundled and shipped together. This service exhibited mutual benefits to both the customer and freight forwarder. This process is the sole source to achieve lower transportation cost per unit weight.

Shipment consolidation can be carried out in four ways:
Inventory Consolidation: This can be achieved by creating the inventory items as per the observed demand. This transforms the large and full size loaded shipments into inventory.

Vehicle Consolidation: This involves merging the pick-up and deliveries of various locations when the vehicle capacity is lower, thus more than one pick-up and delivery are placed on the same vehicle for efficient transportation. Vehicle routing and scheduling employs this type of economy.

Warehouse Consolidation: This consolidation is based on the fundamental principal that the smaller shipments are sent through short distances and larger shipments size are routed through longer distances.

Temporal Consolidation: In this type of consolidation the customers' orders are retained so that larger shipments can be made at one time instead of delivering smaller shipments several times. The benefits from temporal consolidation can be incurred through improved routing and lower per-unit rates. However, the service level might be deteriorated which might result from the failure to ship the deliveries as soon as they are received and filled.

Advantages

Cost Advantage: There are many cost advantages due to freight consolidation, for the customers as well as the shipping company. Consolidations lower overall fuel cost for the freight forwarder. The smaller shipments also lower the total shipping costs for the retailer.
Safety Advantages: It offers safety advantage for forwarding small shipments separately. It reduces the risk of interrupting the delivery of at least one shipment. Hence it lowers the risk level for the shipments.

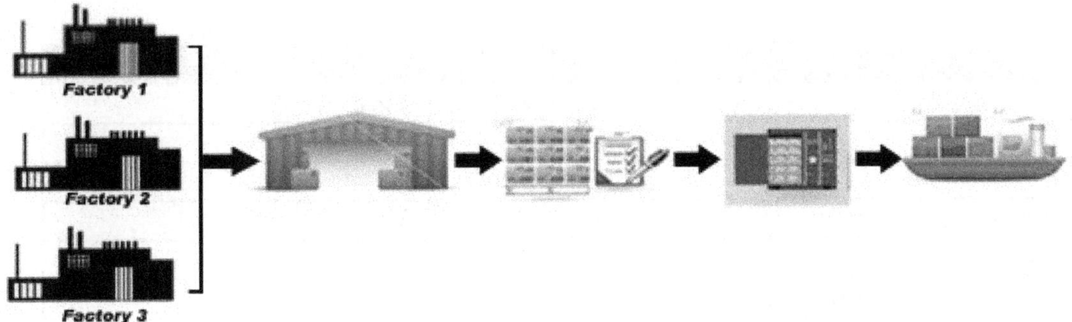

FREIGHT CONSOLIDATION OF CONTAINERS

FIGURE 2

VII. PROBLEMS

1. Suppose we have the following deliveries for the next 3 days

FROM:	DAY 1	DAY 2	DAY 3
HOUSTON			
TO: AUSTIN 4,000 lb.	15,000 lb.	12,000 lb.	
SAN ANTONIO 8,000	20,000	25,000	
DALLAS 32,000	32,000	40,000	

Consider shipping these orders each day or consolidating them into one shipment. Suppose we know the transportation rates

Given rates as quoted by carriers operating in the region, the following costs are incurred if orders are shipped on the day they are received.

Day 1		
	Rate * Volume = Cost	
Austin	6.5 * 40 = 260	
San Antonio	4.5 * 80 = 360	
Dallas	9.3 * 320 = 2976	
TOTAL = 3596 $		

Day 2		
	Rate * Volume = Cost	
Austin	8.5 * 150 = 1275	
San Antonio	7.5 * 200 = 1500	
Dallas	11.3 * 320= 3616	
TOTAL = 6391 $		

Day 3	
	Rate * Volume = Cost
Austin	10.5 * 120 = 1260
San Antonio	14.5 * 250 = 3625
Dallas	9.3 * 300 = 2790
TOTAL = 7625 $	

TOTAL COSTS = 3596 + 6391 + 7625 = 17612

Computing Transport cost for one combined, three day shipment

Day 3	
	Rate * Volume = Cost
Austin	2.5 * 310 = 775
San Antonio	1.9 * 530 = 1007
Dallas	4.3 * 940 = 4042
TOTAL = 5824 $	

The total cost estimated after freight consolidation is lesser than separate shipments.

2. Raw materials are to the shipped from the supplier to the manufacturing site, which is located far away from the supplier. The supplier has taken the responsibility for transportation for the delivery of materials. The traffic manager has three transportation service choices for delivery-rail, water, and truck. He has compiled the following information:

The expected volume to be shipped is 1, 40,000 lb. The product is worth $ 30 per lb. and the plant and carrying costs are 25% per year.

Other data are:

Transport Mode	Transit Time (Days)	Rate $/unit	Shipment Size (Units)
Rail	8	22.00	12,000
Waterways	14	46.00	7,000
Truck	5	95.00	4,000

Which mode of transportation should the supplier select?

RAIL

COST TYPE	COMPUTATION	RAIL	COST
TRANSPORTATION	RD	22(140000)	$ 3,080,000
INTRANSIT INVENTORY	ICDT/365	[.25(30)(1,400,000)(8)]/365	$ 230136
PLANT INVENTORY	ICQ/2	[.25(30)(12000)/2]	$ 45000
WHARE HOUSE INV	ICRQ/2	[.25(30)(22)(12000)/2]	$ 990000
		TOTAL COST	$ 4,345,136

WATERWAYS

COST TYPE	COMPUTATION	WATERWAYS	COST
TRANSPORTATION	RD	46(140000)	$ 6440000
INTRANSIT INVENTORY	ICDT/365	[.25(30)(140000)(14)]/365	$ 40273
PLANT INVENTORY	ICQ/2	[.25(30)(7000)/2]	$ 26250
WHARE HOUSE INV	ICRQ/2	[.25(30)(46)(7000)/2]	$ 1207500
		TOTAL COST	$ 7714023

TRUCK

COST TYPE	COMPUTATION	TRUCK	COST
TRANSPORTATION	RD	95 (140000)	$ 13300000
INTRANSIT INVENTORY	ICDT/365	[.25(30)(140000)(5)]/365	$ 14383
PLANT INVENTORY	ICQ/2	[.25(30)(4000)/2]	$ 312500
WHARE HOUSE INV	ICRQ/2	[.25(30)(95)(4000)/2]	$ 1425000

In cost point of view the supplier would select truck mode of transportation. However, upon consideration of time depending on the delivery rate he would choose truck for faster delivery.

3. CARRIER ROUTING

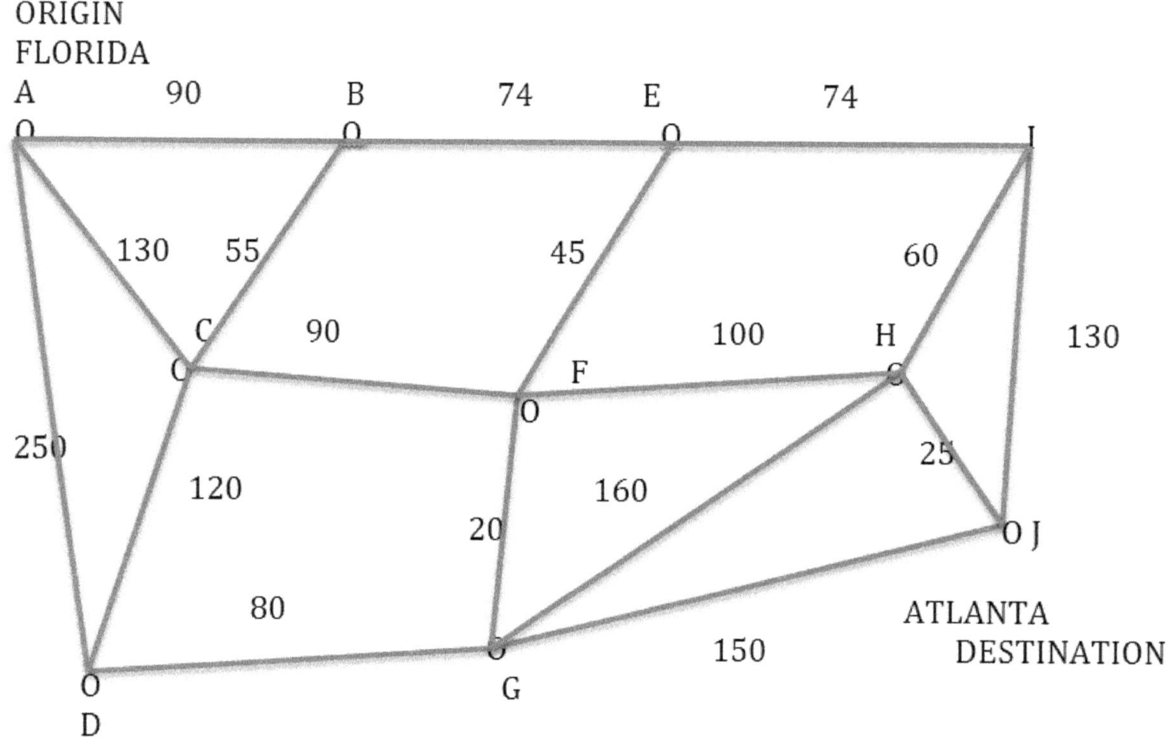

FIGURE 3

STEP	SOLVED NODES DIRECTLY CONNECTED TO UNSOLVED NODES	ITS CLOSEST CONNECTED UNSOLVED NODES	TOTAL COST INVOLVED	nth NEAREST NODE	Its Minimum Costs	Its
I	A	B	90	B	90	AB
2	A	C	130	C	130	AC
	B	C	90+55=145			
3	A	D	250			
	B	E	90+74=164	E	164	BE
	C	F	130+90=220			
4	A	D	250	F	220	CF
	C	F	130+90=220			
	E	I	90+74+74=238			
5	A	D	250	I	238	EI
	F	H	130+90+100=320			
	E	I	90+74+74=238			
6	A	D	250			
	F	H	130+90+100=320	D	250	AD
	I	H	90+74+74+60=298			
7	F	H	130+90+100=320	H	298	IH
	G	J	250+80+150=480			
	I	H	90+74+74+60=298			
8	H	J	90+74+74+60+25=323	J	323	HJ
	G	J	250+80+150=480			
	I	J	90+74+74+130=368			

4. The Varun motors Trucking company uses vans to pick up merchandise from outlying customers. The merchandise is returned to a depot point, where it is consolidated into large loads to the moved over long distances. A Days pickups are shown in the figure below. How should the routes be routes be designed for minimum total travel distance?

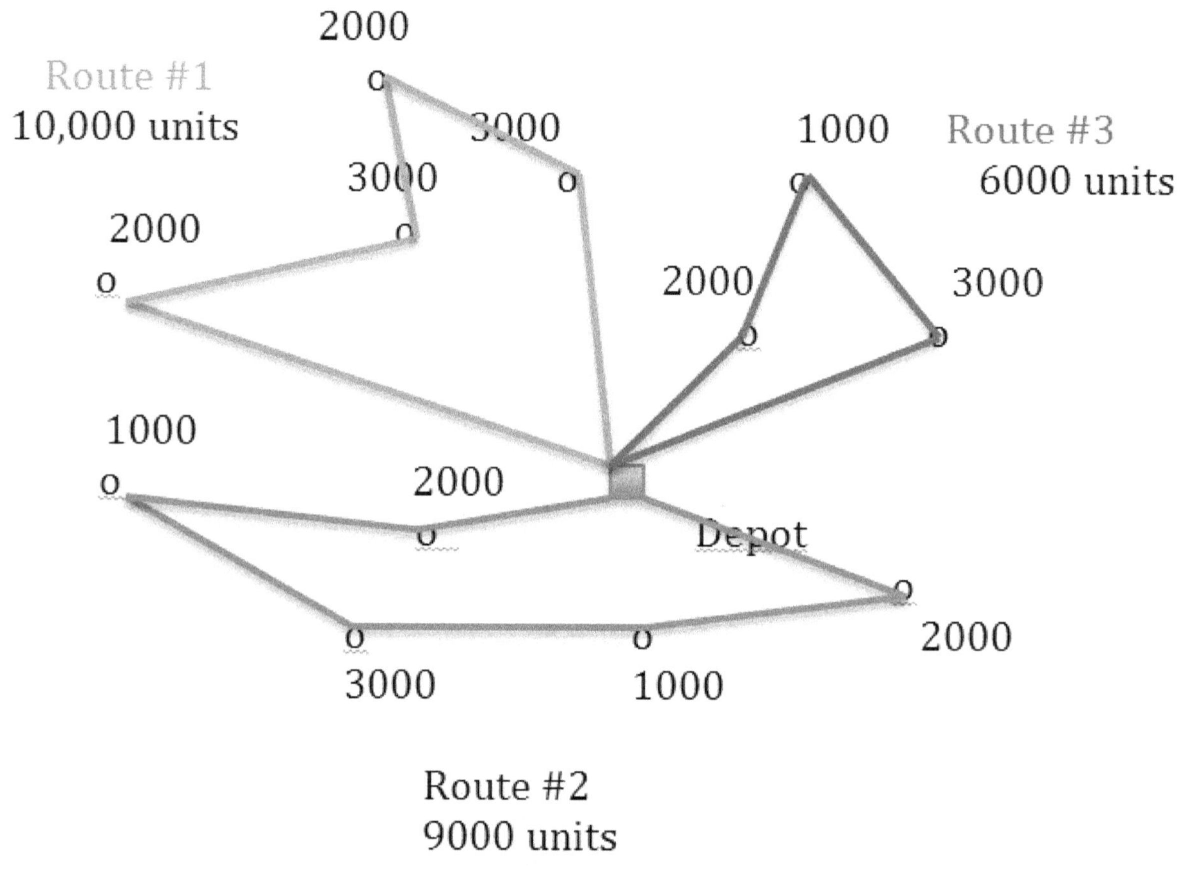

FIGURE 4

REFERENCES

1. Business Logistics/Supply Chain Management, Ronald H. Ballou, Samir K. Srivastasa
2. http://eptrend.com/more-services/container-consolidation/
3. http://www.ornl.gov/sci/gist/TRAGIS_2005.pdf

LECTURE 12
READING MATERIALS

12.0 Introduction

The transportation infrastructure is necessary for our nation's security and is a very important part of our military. The military relies on modes of transportation that remain secure. A secure transportation system is a "must have" in order to combat terrorist and their activities. If our transportation becomes vulnerable we lose capabilities to move troops and equipment to desired locations. A large portion of the economy is connected with transportation assets, which relies on transportation that is functional and reliable. The disruption of any transportation mode can pose undo harm on the supply chain. This is why transportation, which ever mode you want to choose, is a target when anyone wants to slow the economy. There is historical data that shows attacks on concentrated areas, and transportation hubs as available targets. All modes of transportation are targets, and usually the high profile locations are more targeted and required upgraded security.

People rely on transportation and it is a necessity. It is relied on for daily transport and use for maneuvering equipment in a theater of war, and the transport of civilians to achieve economic development. The military counts and depends on all modes of transportation daily to move equipment, supplies and personnel in order to maintain readiness. Transportation is an economic engine, whether it is for the military or commercial. Transportation is a means of security for the people of our country. The security of transportation, with the flexibility and available resources is critical to maneuverability.

12.1 Definition

Transportation defined:

- ➢ carry or move from one point to another
- ➢ transport by rail, truck or bus
- ➢ a ship or plane employed for transporting Soldiers and military equipment
- ➢ aircraft carrying freight or passengers as part of a transportation system.

Security defined:

A condition that results from the establishment and maintenance of protective measures that ensures a state of inviolability from hostile acts or influences - Dictionary of Military and Associated Terms.

Military security is associated with a countries defense and to deter others from aggression towards the military and its nation. The term "military security" is considered synonymous with "security" in much of its usage. One of the definitions of security given in the *Dictionary of Military and Associated Terms*, may be considered a definition of "military security" "Security." *in* "Dictionary of Military and Associated Terms", 2001 (As amended through 31 July 2010) op.cited. Pg 477. Accessed 26 September 2010).

12.1.2 Importance of Transportation and Security

The security of transportation is recognized as one of the earliest forms of national security. Through the years, the scope of military security has expanded from conventional forms of conflict. We have been involved in warfare, in the most recent years, from Kuwait, Irag an Afghanistan. Paleri, Prabhakaran (2008). *National Security: Imperatives And Challenges.* New Delhi: Tata McGraw-Hill. p. 521. ISBN 978-0-07-065686-4. Retrieved 23 Septamvber 2010.

12.2 Modes of Transportation

- Air: rotary wing (helicopters) transport or cargo, fixed wing (aircraft) transport or cargo

- Rail: cargo trains carrying military equipment usually to a port

- Road: trucks and tractor/trailer rigs delivering to port or other destinations within a specific area of operation

- Water: cargo vessels (ships/barges) that carry or haul rolling stock, track vehicles, helicopters, supplies, containers and a large contingent of pre-staged supplies and equipment.

12.2.1 Air

Strategic transport for the military consists of aircraft along with other modes. Cargo aircraft are typically fixed wing with additional support from the rotary wing cargo aircraft. Rotary wing aircraft is usually utilized to deliver troops and equipment by an array of methods. This is usually done outside of the commercial flight routes and in uncontrolled airspace. Fixed wing aircraft are utilized to transport and deliver airborne units, transport aircraft, and have multi-roles and duties of aerial refueling. Some military transport aircraft are tasked to perform multi-role duties such as aerial refueling, tactical, operational and strategic airlifts onto runways that are constructed in a combat area by engineers or an improved runway that has been maintained for continuous operations.

Military transport helicopters are commonly utilized in locations where conventional aircraft use is difficult. A transport helicopter is the primary transport for many light units that require immediate insertion at unconventional locations. When a helicopter is unable to touch ground, it can hover for a period of time and allow troops to descend to the earth by the use of ropes and other various methods.

Transport helicopters are identified by: assault, medium haul or heavy haul. Assault helicopters are the lightest of transport types, and are designed to transport small units quickly and effectively. Helicopters in the assault role are generally armed for self-protection both in transit and for suppression of the landing zone. A utility helicopter is one that is usually utilized for medium lift capabilities. The medium transport helicopter is the most common used during troop and cargo movement. A medium transport can move up to a platoon of infantrymen, and can carry the equipment needed for support.

Heavy lift helicopters are the largest and most capable of the transport types, these helicopters operate in a tactical transport role in much the same way as small fixed wing turboprop air-lifters. The lower speed, range and increased fuel consumption of helicopters are more than compensated by their ability to operate virtually anywhere.

Kazak Mi-17 helicopter   CH-54B carrying an M551 Sheridan tank

C-17 Globemaster III military cargo aircraft

12.2.2 Rail

The U.S. Military Railroad (USMRR) was established by the United States War Department as a separate agency to operate any rail lines seized by the government during the American Civil War. The United States Congress authorized President Abraham Lincoln to seize control of the railroads and telegraph for military use in January 1862. In practice, however, the USMRR restricted its authority to Southern rail lines captured in the course of the war. As a separate organization for rail transportation the USMRR is one of the predecessors of the modern United States Army Transportation Corps (Gable, *Railroad Generalship*, p. 13).

Military rail load during a winter storm, spotters and ground guides assisting the load of equipment

12.2.3 Road

Mine-Resistant Ambush Protected (MRAP; pron.: /ˈɛ mræp/ *EM-rap*) vehicles are armored fighting vehicles used by various armed forces, whose designed purpose is surviving improvised explosive device (IED) attacks and ambushes. The first development in armored vehicles designed specifically to counter the land mine threat were initiated during the Rhodesian Bush War; existing technology was subsequently inherited (and matured) by the South African Defence Force after 1980 (Russell, Robert W (2009).

There are many vendors that design the MRAP, and each has its own unique design and each with its own vehicle or family of vehicles. Leaders within the military have pushed and replaced a lot of vehicles with MRAP vehicles. Armored vehicles were considered a need that was of the upmost urgency in Afghanistan, and the program to implement was funded under and emergency war budget May 2007. The dollar amount associated with this program was 1.1 billion dollars (*Defense Industry Daily*. 25 Apr 2007).

In June 2008, it was reported that attacks had decreased by ninety percent and were partially contributed to the MRAP. Without a doubt the MRAP has saved numerous lives of troops during the last decades. The fourteen ton MRAP has forced insurgents to build bigger, more sophisticated bombs to disable the vehicles. Building bombs/IED large enough to disable a MRAP takes additional time and resources compared to IED to destroy conventional wheeled vehicles. (Gannett Government Media Corporation. Retrieved 9 May 2011). This program is very similar to the United States Army's Medium Mine Protected Vehicle program. US Army: 17,000 MRAP Vehicles to Replace HMMWV. *Defense Industry Daily*. 11 May 2007

MRAP Wheeled Vehicle

MRAP Wheeled Vehicle

12.2.4 Water

There are multiple aircraft carriers that are deployed around the world, and are able to place the military's air power within striking distance. These aircraft carriers maintain a critical role in our nation's security. There are joint operations that are carried out as exercises and named operations, along with the leaders of the nation relying on the nearest aircraft carrier located near the conflict. Again, these aircraft carriers are the first line of defense for our nation. Former President Bill Clinton summed up the importance of the aircraft carrier by stating that "when word of crisis breaks out in Washington, it's no accident the first question that comes to everyone's lips is: where is the nearest carrier?" These aircraft carriers are made up of rotary and fixed wing assets, and these carriers protect friendly forces, conduct electronic warfare, assist in special operations, and carry out search and rescue missions. The carriers also serve as command centers for large forces or multinational task forces. The U.S. Navy aircraft carriers can also host aircraft from other nations' navies; the French Navy's Rafale has operated, during naval exercises, from U.S. Navy flight decks. This is an example on how the military and the civilian agencies use like methods of conducting business. Both operate jointly with other countries and organizations along with maintaining network capabilities.

12.2.5 Multimodal

The transportation of cargo with at least two different modes of transport, while utilizing the same contract is referred to a multimodal. The carrier remains responsible for the entire shipment (example: road, rail, sea or air). The carrier may not have all of the modes of transport under his umbrella; however the carrier is still responsible.

Article 1.1 of the United Nations Multimodal Convention defines multimodal transport as follows: "'International multimodal transport' means the carriage of goods by at least two different modes of transport on the basis of a multimodal transport contract from a place in one country at which the goods are taken in charge by the multimodal transport operator to a place designated for delivery situated in a different country". (United Nations Convention on International Multimodal Transport of Goods (Geneva, 24 May 1980). During 2011 to present multimodal transport "containerized" is the most important part of the multimodal transport.   It is also important to note that multimodal is not equivalent to container transport. Multimodal transport is feasible without any form of container being involved.

12.2.6 Intermodal

Intermodal freight requires freight to move in a container or vehicle utilizing multiple modes of transportations: rail, ship and truck without freight being handled (trans-load, etc.) Intermodal reduces the handling and improves security, as well as loss or damage of freight.   This is a conduit for freight to be moved quickly to the end destination. The reduced costs utilizing over the road trucking is the key benefit for intercontinental use, as well as reduced greenhouse gas emissions. Time is also reduced greatly (http://www.mjc2.com/e-freight-logistics.htm Intermodal Optimization).

This transportation goes back into history around the 18th century. The earliest containers were used for shipping coal. Coal containers were referred to as "loose boxes" and deployed on canals and railways and were used for road/rail transfers. Wooden coal containers were used on railways dating back to the 1830s (Liverpool and Manchester Railway). It was not until the early 1900s that the first covered container was adopted, and its main purpose was to move furniture.

When the three links of collection is combined, the transfer and delivery in multimodal transportation, a directional multi-phase labeling algorithm is proposed to show the shortest path cost problem in container, motor rail, multimodal transportation, with reference to the Dijkstra labeling algorithm. The concept of the railway container freight station choice set is introduced in, railways and highways are differentiated, and delivery costs at container freight stations are included. Several computing examples are given. The designed computation program indicates that the algorithm is effective and practicable. **HE Guo-xian (School of Traffic and Transportation, Lanzhou Jiao tong University, Lanzhou 730070, China).**

Selection of appropriate intermodal routes required procedures for linking freight origins and destinations to the transportation network, procedures for modeling intermodal terminal transfers and inter-carrier interlining practices, and a procedure for generating multimodal impedance functions to reflect the relative costs of alternative, survey reported mode sequences. (Transportation Research Part C: Emerging Technologies, Volume 8, Issues 1–6, February–December 2000, Pages 147–166).

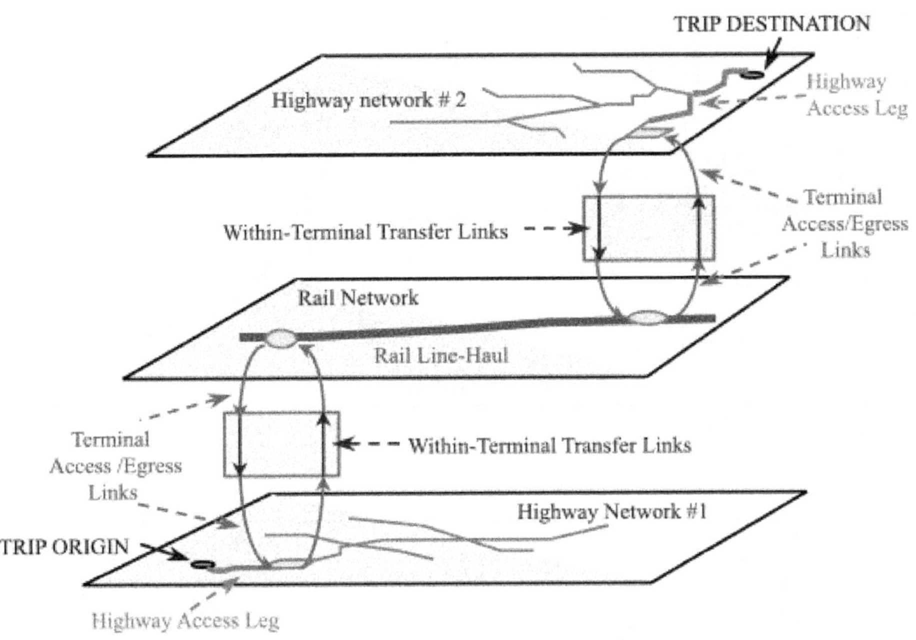

Fig. 1. Construction of a multi-layer intermodal shipment routing.

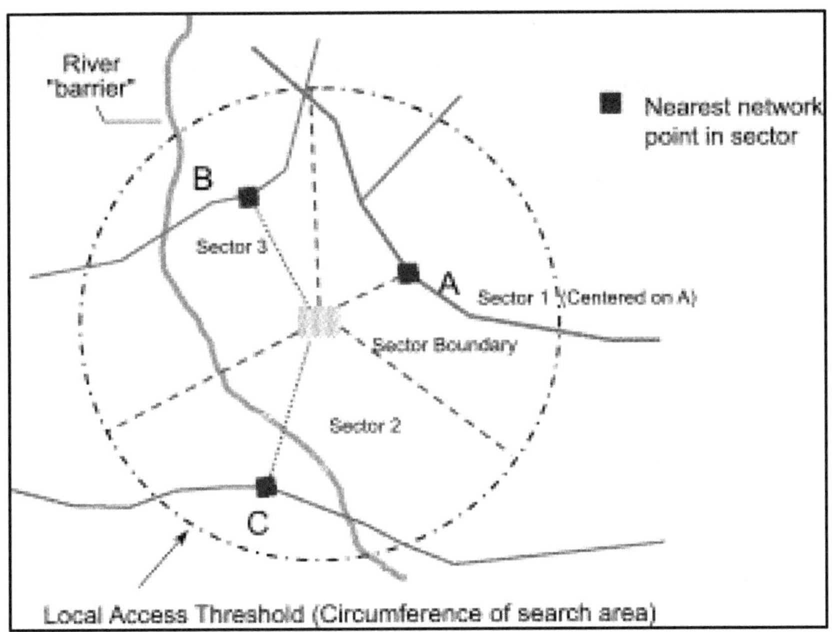

Fig. 2. Highway access modeling using distance threshold and sector search.

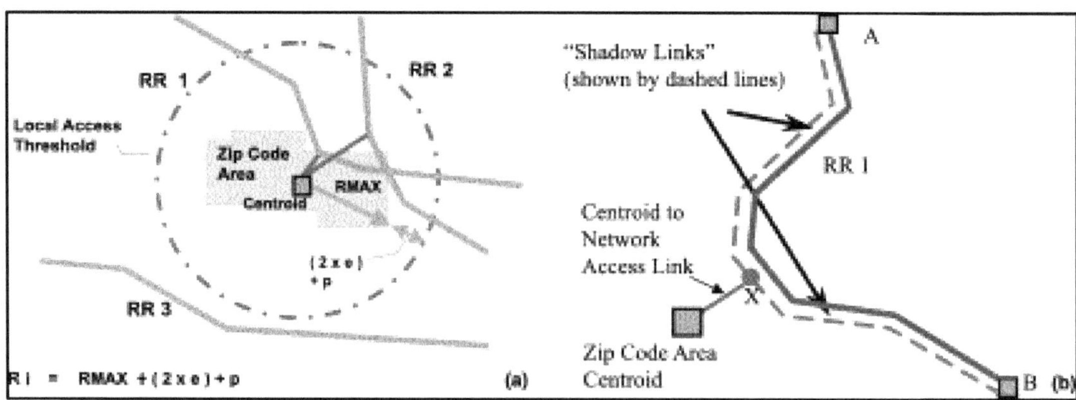

Fig. 3. Rail line access modeling.

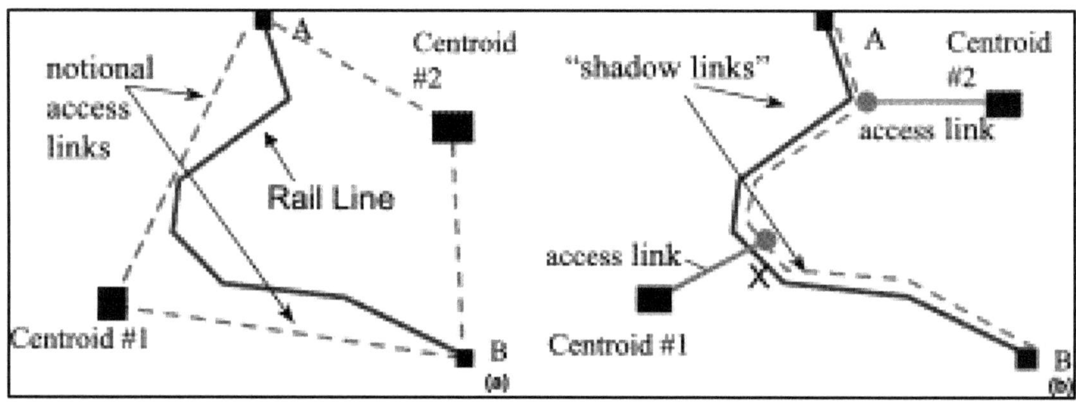

Fig. 4. Alternative rail network connection methods.

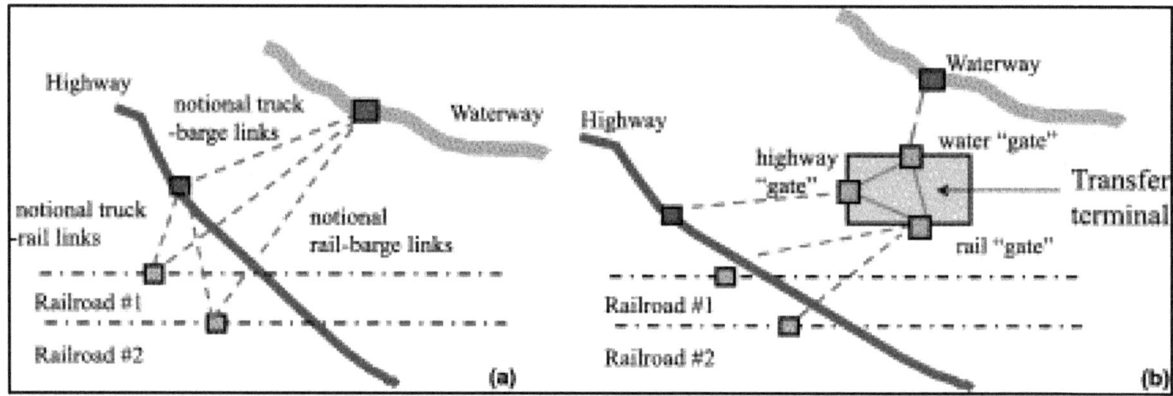

Fig. 5. Two ways to model intermodal terminal transfers.

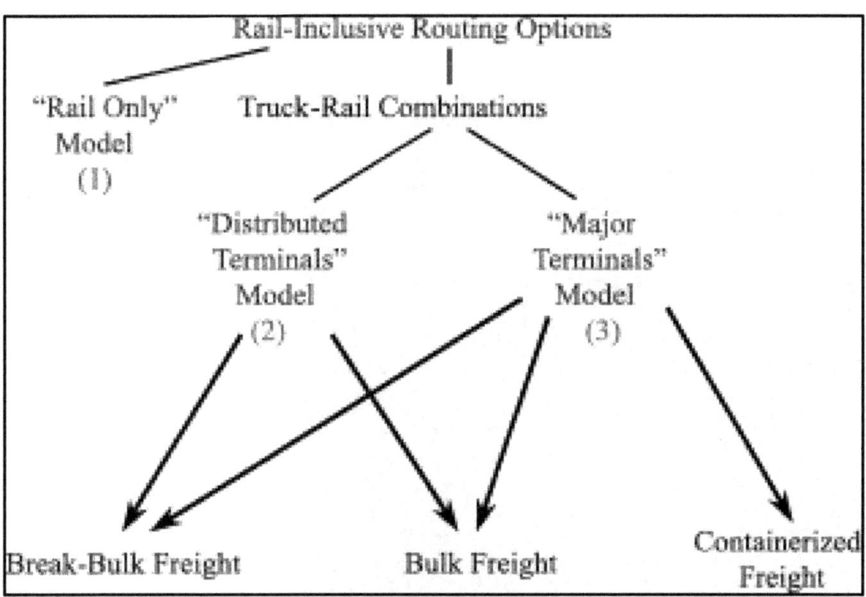

Fig. 6. Family of rail-inclusive freight shipment routing models.

12.3 Military and Civilian Transportation

In the commercial sector, many Fortune 500 corporations have used a tool called the Supply Chain Operations Reference (SCOR) model to decrease costs and increase profits, and this will improve strategic competitiveness. DOD 4140.1R, DOD Supply Chain Material Management Regulation, directed that "DOD Components shall use the supply chain operational reference processes of Plan, Source, Maintain/Make, Deliver, and Return as a framework for developing, improving, and conducting material management activities." Logisticians will gather knowledge and understand the SCOR model.

12.3.1 Similarities

The strategic level of our military along with the private agencies or sector, have very similar constraints and limitations. The logistics concern of distribution has to be addressed. Adopting blanket solutions without vetting will not be a wise move, however with enough forethought and trial implementation there is much success to be achieved. Many of the problems faced in today's DOD supply chain are the same ones that the commercial sectors share or have dealt with in the past.

By adopting some of the best business practices, and refine them to fit the military and using a SCOR model can improve the efficiency and effectiveness of the Department of Defense supply chain.

Soldiers from the 626th Brigade Support Battalion, 3d Brigade Combat Team, 101st Airborne Division, load a box of supplies onto a truck at Forward Operating Base Dragon in Iraq. The war fighter is the customer at the end of the DOD supply chain. (Photo by TSgt. Adrian Cadiz, USAF.)

12.3.2 Differences

Effective supply chain management originates from understanding the needs of the customer at the end of the supply chain (the war fighter). It requires developing and establishing metrics that are not actually seen or heard of in the commercial sector. The SCOR model provides a view at the entire supply chain to determine how to best meet the war fighter's requirements. Logisticians gather critical data on a daily basis to make sure the war fighter's needs are being meet. Military logisticians are measured by order backlog, order fill time, and days of inventory stock, that, when measured against metrics, can provide the means to improve performance. When everyone understands the metrics used it allows better coordination. Understanding this approach or concept can lead to reduced costs and better performance.

Supply chain management will integrate critical business steps outlined by the SCOR model. Do not overestimate the time and work that is allowed to create a seamless organization across the Department of Defense, the Defense Logistics Agency, the U.S. Transportation Command, and many other military services. The SCOR model and best practices will increase effectiveness and efficiencies

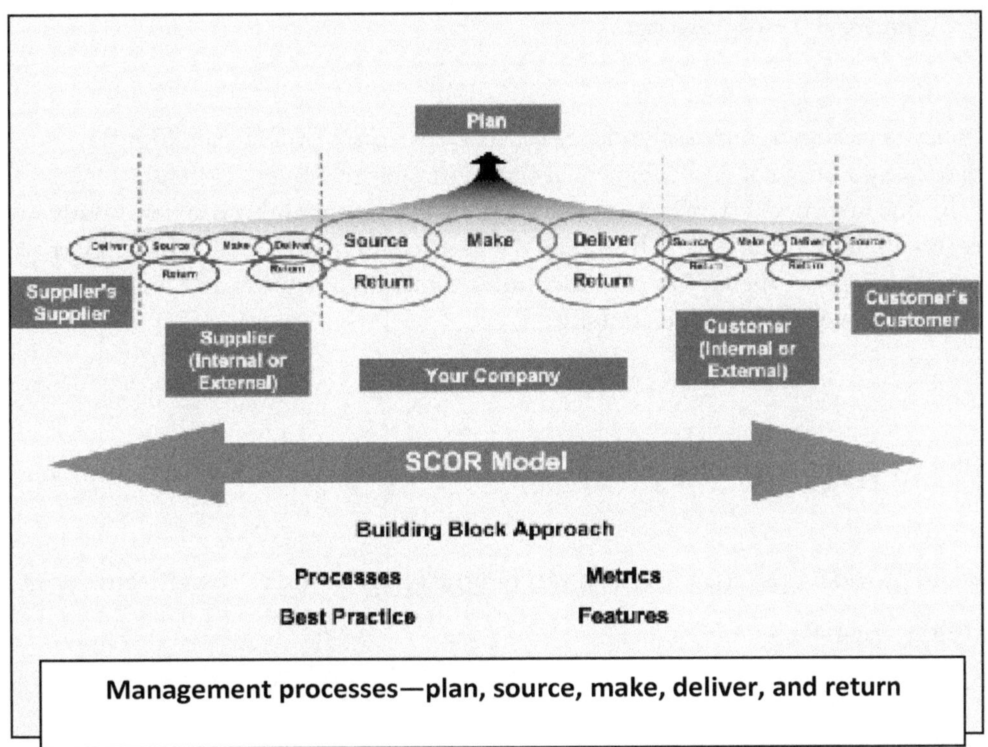

Management processes—plan, source, make, deliver, and return

What is the difference between military aircraft and civil aircraft?

Military aircraft are designed to destroy the enemy's military aircraft and (sometimes) their civil aircraft. They also perform bombing tasks as well as a few non-combat tasks. Civil aircraft are mainly transports and crop sprayers that don't do much fighting.

Military aircraft are designed to fulfill a multitude of roles that are specified by the Armed forces that will be using these aircraft. These roles include Air attack as well as air support and defense of other military assets. Transport and surveillance are also key roles that military aircraft fulfill. Civil aircraft are designed to perform commercial transport as well as personal use and sports roles. Civil aircraft have no need for stealth or combat survivability the way military aircraft require these characteristics.

12.3.3 Coordination with Commercial Transportation

Transportation Policy is created to provide guidance to the Department of Defense (DoD) Components for effective use of government and commercial transportation resources. Defense Transportation is top of the line intermodal transportation system. It is efficient and is partnered with commercial agencies to ensure military readiness. The military is able to sustain itself with the support of commercial providers, as well as improve the life support function for military members and their families. This stands true on deployments or based stateside.

Functions

- Serves as Principal Advisor for establishing policies and providing guidance to the Department of Defense

- Develops policies and provides analysis, advice and recommendations to ensure cost-effective joint logistics support to the war-fighter

- Coordinates and negotiates with federal departments and inter-agencies

- Ensures that the Defense Transportation is effective in providing "end-to-end" support to the war-fighter

- Establishes and maintains Department of Defense transportation, distribution, traffic management, and strategic mobility policies

- Collaborates with the Distribution Process Owner and its distribution partners to support distribution and process improvement initiatives

- Promotes coordination, cooperation, and mutual understanding between Department of Defense

- Provides oversight and guidance for the development and implementation of Department of Defense automation programs (http://www.acq.osd.mil/log/tp/ (Deputy Assistant Secretary of Defense (Transportation Policy) DASD (Transportation Policy)

12.4 Transportation Security

Border and transportation security is a critical part of the Department of Homeland Security's (DHS) national strategy. DHS strategy calls for the creation of "smart borders" where information from local, state, federal, and international sources can be combined to support risk-based management tools for border-management agencies. This paper proposes a framework for effectively integrating such data to create cross-jurisdictional criminal activity networks (CAN)s. Using the approach outlined in the framework, we created a CAN system as part of the DHS-funded BorderSafe project. This paper describes the system, reports on feedback received from investigating officers, and highlights key issues and challenges. Marshall, B., Kaza, S., Xu, J., Atabakhsh, H., Petersen, T., Violette, C., & Chen, H. (2004, October). Cross-jurisdictional criminal activity networks to support border and transportation security. In

Intelligent Transportation Systems, 2004. Proceedings. The 7th International IEEE Conference on (pp. 100-105). IEEE.

12.4.1 Security Risk

Security risk is expressed as a mathematical function. Security risk is a product of probability of attempt on targets, vulnerability of the target, and the damage costs when the security is breached.

Security Risk = Attempt on target x Probability of incident x Vulnerability x Damage

This equation suggests the nature of the risk to the transportation sector. Through the years domestic security concerns have been minimal, however the last 12 years have change our approach. The probability of an incident attempt 12 years ago was to be hardly heard of in this nation. Now the magnitude of an event is detrimental to the nation, the attacks are sophisticated and hard to see coming. The security breaches are planned and usually in strategic locations, that will impede the response time for a counterattack. The explosives are more powerful, strategically placed, and create a higher awareness for both the public, private and military sectors.

While the above calculation could be applied to individual services and facilities, it can also be applied at the systems level where it would suggest that the security risk is now far greater. We have implemented heightened security by planning and design. Heightened security impacts on the way transportation is planned and conducted. Our common ways of self-moving or utilizing public transportation will change as actions for increased security is implemented.

12.4.2 The Impacts of Security Concerns on Transportation

After 10 years of transportation security, do we need to rethink our current actions for airline check-ins or rapid movement on rail transit? Is the ease of access becoming a complacent area for us to operate in, are we guarded or are we on the right track? The use of information technology is a great thing; however it can be compromised by the best. Are there certain modes of transportation that are more attractive for the security to be compromised? It is of great concern that the public is willing to consider placing more investments in transportation systems that provide security. Security initiatives are in competition with other competing agencies for funding; however these investments can impact transportation services over time. The physical locations of transportation resources are targeted more often, and various security threats impact transportation facilities and their services.

Table I: Scenarios Considered in the U.S. DOT Vulnerability Assessment

Physical Attacks

• Car bomb at bridge approach	• Attack on passenger vessel in port
• Series of small explosives on highway bridge	• Shooting in rail station
• Single small explosive on highway bridge	• Vehicle bomb adjacent to rail station
• Single small explosive in highway tunnel	• Bombing of airport transit station
• Car bomb in highway tunnel	• Bombing of underwater transit tunnel
• Series of car bombs on adjacent bridges or tunnels	• Bus bombing
• Bomb(s) detonated at pipeline compressor stations	• Deliberate blocking of highway-rail grade crossing
• Bomb detonated at pipeline storage facility	• Terrorist bombing of rail tunnel
• Bomb detonated on pipeline segment	• Bomb detonated on train in rail station
• Simultaneous attacks on ports	• Vandalism of track structure and signal system
• Terrorist bombing of waterfront pavilion	• Terrorist bombing of rail bridge
• Container vessel fire at marine terminal	• Explosives attack on multiple rail bridges
• Ramming of railroad bridge by maritime vessel	• Explosive in cargo of passenger aircraft

Biological Attacks

• Biological release in multiple subway stations	• Anthrax release in transit station
• Anthrax release from freight ship	• Anthrax release on passenger train

Chemical Attacks

• Sarin release in multiple subway stations	• Physical attack on railcar carrying toxics

Cyber and C3 Attacks

• Cyber attack on highway traffic control system	• Sabotage of train control system
• Cyber attack on pipeline control system	• Tampering with rail signals
• Attack on port power/telecommunications	• Cyber attack on train control center

Source: National Research Council, *Improving Surface Transportation Security, A Research and Development Strategy*, Washington D.C: National Academy Press, 1999.

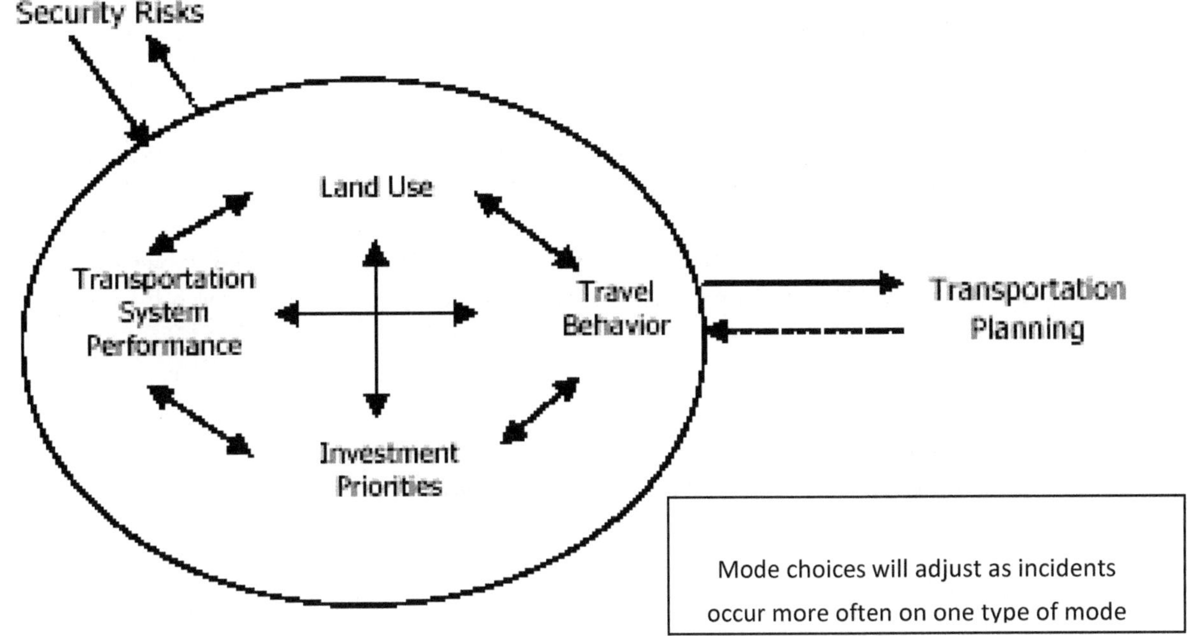

12.5 Conclusion

Security considerations will control on how transportation is implemented. Planning processes, analytical tools, structures of organizations will facilitate change due to security concerns. Transportation has always been in the forefront in regards to security risk. Our long range plans and ways of thinking will change as plans are refined. Transportation planners will need to use creative thinking, think long range, and try to look into the future and anticipate. Cost is a concern and must be dealt with, however our nation's security does not have a price tag.

References:

Davidson, A. L. (2006). *Key performance indicators in humanitarian logistics* (Doctoral dissertation, Massachusetts Institute of Technology).

"Dictionary of Military and Associated Terms", 2001 (As amended through 31 July 2010) op.cited. Pg 477. Accessed 26 September 2010

Dan Lamothe (9). "New weapons, war dogs eyed to fight IEDs". Gannett Government Media Corporation. Retrieved 9 May 2011.

Flynn, Stephen E. "Transportation Security: Agenda for the 21stCentury." *TR News* 211, November-December 2000.http://www.nas.edu/trb/publications/security/sflynn.pdf (2002).

Gable, *Railroad Generalship*, p. 13.

"Global Intermodal Freight: State of readiness for the 21stCentury: Report of a Conference; February 23-26, 2000; Long Beach, California." Transportation Research Board

HE Guo-xian(School of Traffic and Transportation,Lanzhou Jiaotong University,Lanzhou 730070,China)

http://en.wikipedia.org/wiki/Wikipedia:Citing_sources

http://www.boeing.com/stories/videos/vid_16_apache.html?cm_ven=Paid+Search+Google&cm_cat=Innovation&cm_pla=Military&cm_ite=Aircraft+Apache

http://www.mjc2.com/e-freight-logistics.htm Intermodal Optimization

http://www.acq.osd.mil/log/tp/

Marshall, B., Kaza, S., Xu, J., Atabakhsh, H., Petersen, T., Violette, C., & Chen, H. (2004, October). Cross-jurisdictional criminal activity networks to support border and transportation security. In *Intelligent Transportation Systems, 2004. Proceedings. The 7th International IEEE Conference on* (pp. 100-105). IEEE.

National Research Council, 2001. http://www.nas.edu/trb/publications/security/cp25.pdf(2002).

National Research Council, *Improving Surface Transportation Security, A Research and Development Strategy*, Washington D.C: National Academy Press, 1999; originally in U.S. DOT, *Surface Transportation Vulnerability Assessment*, Final Report, Washington D.C. May, 1998.

O'Neil, Daniel J. "Statewide Critical Infrastructure Protection: New Mexico's Model." *TR News* 211, November-December 2000. http://www.nas.edu/trb/publications/security/doneil.pdf(2002).

Paleri, Prabhakaran (2008). *National Security: Imperatives And Challenges*. New Delhi: Tata McGraw-Hill. p. 521. ISBN 978-0-07-065686-4. Retrieved 23 September 2010.

Polzin, Steven E. "Transportation Planning After September 11th, 2001." *The Urban Transportation Monitor,* December 7, 2001.

Qunming, L., Guoning, S., & Shilian, Z. (2003). Performance Measuring Indicators of Supply Chains. *China Mechanical Engineering, 10,* 021.

Roadside bombs decline in Iraq - USATODAY.com

Russell, Robert W (2009). *Does the MRAP meet the U.S. Army's needs as the primary method of protecting troops from the IED threat?* (Master of Military Art and Science thesis). US Army Command and General Staff College..

"Security." *in* "Dictionary of Military and Associated Terms", 2001 (As amended through 31 July 2010) op.cited. Pg 477. Accessed 26 September 2010

Transportation Research Part C: Emerging Technologies Volume 8, Issues 1–6, February–December 2000, Pages 147–166

United Nations Convention on International Multimodal Transport of Goods (Geneva, 24 May 1980)

US Army: 17,000 MRAP Vehicles to Replace Hummers? *Defense Industry Daily*. 11 May 2007